Lynne Glover
1979

THE MAN WHO WALKED AROUND THE WORLD

Also by Clinton Trowbridge

THE CROW ISLAND JOURNAL

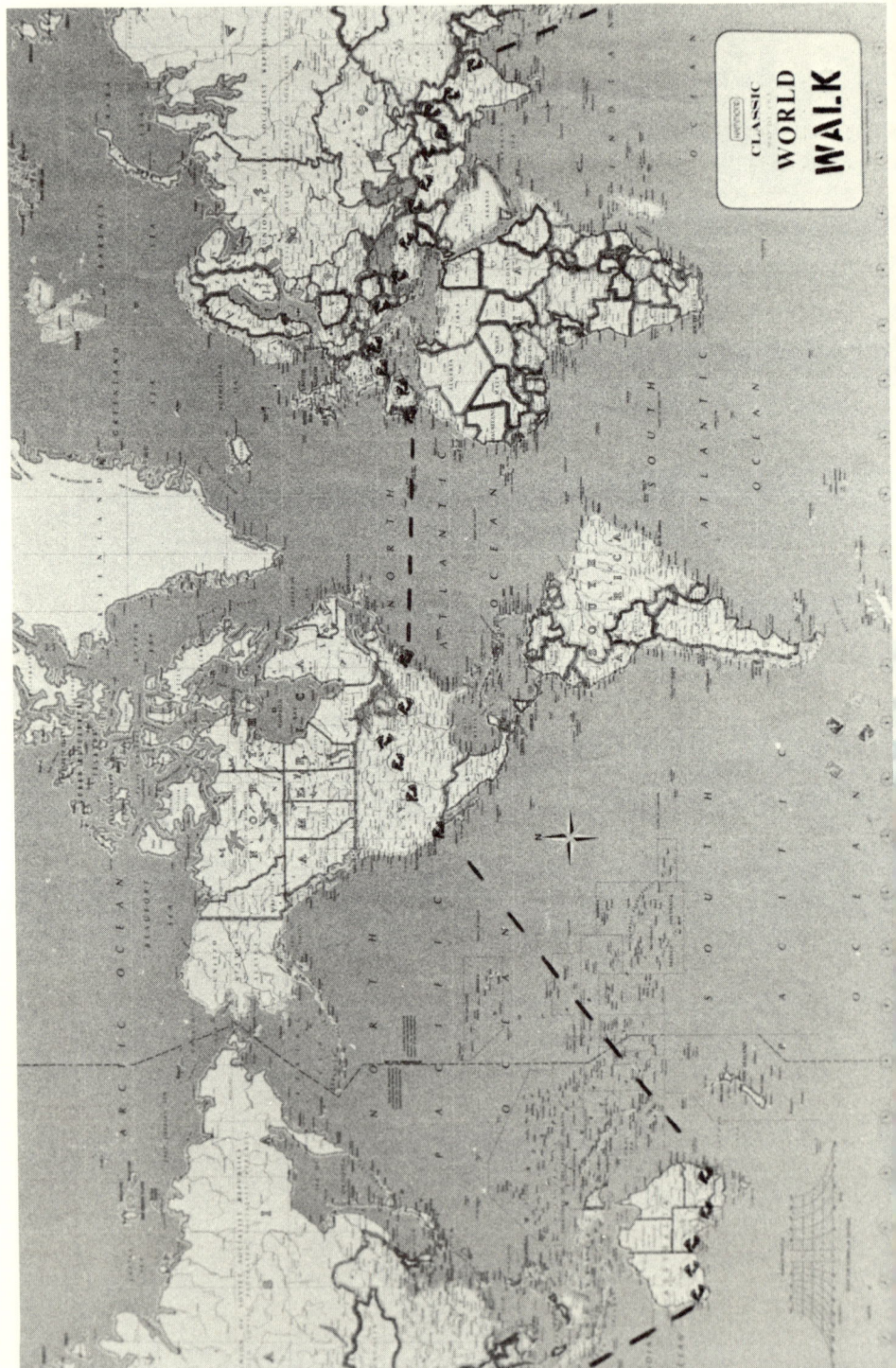

THE MAN WHO WALKED AROUND THE WORLD

A TRUE STORY

by
DAVID KUNST and
CLINTON TROWBRIDGE

WILLIAM MORROW AND COMPANY, INC.
NEW YORK   1979

Copyright © 1979 by David Kunst and Clinton Trowbridge

The article printed from the *Minneapolis Tribune* issue of September 22, 1974, is reprinted with permission of the *Minneapolis Tribune*.

All rights reserved. No part of this book may be reproduced or utilized in any form or by any means, electronic or mechanical, including photocopying, recording or by any information storage and retrieval system, without permission in writing from the Publisher. Inquiries should be addressed to William Morrow and Company, Inc., 105 Madison Ave., New York, N. Y. 10016.

Library of Congress Cataloging in Publication Data

Kunst, David.
    The man who walked around the world.

    1. Kunst, David.   2. Voyages around the world.
I. Trowbridge, Clinton, 1928-   II. Title.
G440.K96K86   910′.92′4   [B]   78-27279
ISBN 0-688-03437-3

BOOK DESIGN   **CARL WEISS**

Printed in the United States of America.

First Edition

1 2 3 4 5 6 7 8 9 10

To my friend Rich, who gave me the idea,

my brother John, who was willing to step

out of the someday, someday world of dreamers

to go along,

and my brother Pete, who took John's place

THE MAN WHO WALKED AROUND THE WORLD

CHAPTER 1

ETCHED IN THE MOONLIGHT, THE SHADOWY, TURBANED FIGURES seem to glide toward us. I see no weapons, so I cannot shoot. They move without a sound, shoulder to shoulder, looking straight at us.

"*Buro!* Go!" I yell, and a shadow moves off to our right.

"*Buro!*" I shout and my finger squeezes the trigger.

Boom! The blast of the 12-gauge crashes off the wall of the gorge. They stop. The figure on our right slowly rejoins the others. I break and reload, pointing the gun at them this time. I dare them to take another step or to show any sign of a weapon. There is a pause—utter silence—and then slowly, very slowly, they start to walk backward, turning slightly after twenty or thirty yards, until they melt back into the night. The whole time they have not made a sound or taken their eyes off us.

I hear John sigh with relief behind me. "I guess the police captain was right," he says.

"Maybe," I say. "But I could have dropped all six of them back there." Suddenly, a shot.

"Run for it!" cries John. I have turned, started to run toward the wagon. Another shot rings out. I am hit in the chest. As I fall, I yell out to John to take the gun.

"This is stupid, you dummies. Please," John says. And

then there is another shot. I hear a groan and at the same time the blast of the shotgun.

"Play dead!" I yell. I am lying on my stomach and though I know I've been shot, I do not believe it. There is no pain, only the shock, and a sensation of numbness. I hear footsteps approaching so I concentrate on lying very still and loose, and on keeping my eyes closed. I am grabbed by the right shoulder and roughly turned over. A hand on my chest, where the bullet has hit me. Suddenly, there is the muffled blast of another shot, and I know then they have finished off John. The urge to feel my thumbs dig into the hairy throat above me is like a great wave pushing me, pushing me upward. But I will not give in to the sweetness of this desire. Instead, I lie there as limply as possible, holding my breath, certain that the pounding of my heart must give me away, certain that it is only a matter of seconds before the gun goes off again. A hand takes hold of my left arm and another slips the watch off my wrist. I let my arm fall to the ground heavily—as if it belongs to someone else. Two hands grab me by the belt and I am half-lifted off the ground. My belt, with its sheath knife, is jerked away, and I am dropped to the earth with a thud. Almost immediately, hands grab my wrists and I am dragged off. Stories of sexual mutilation flash through my mind. Or will they decapitate me with my own knife?

My arms are dropped. Something bulky but soft is thrown on top of me: the mattress from our wagon. I hear them going through our things. The temptation to take just one look is overpowering. I open my right eye and immediately close it again. There is a shoe with a curled toe a foot from my face; the butt of our shotgun is resting next to it. I lie there, for hours it seems; all around me the crash of pots and pans, the thud of cans upon the ground, the grunting and shouting of the Afghans. Yet all of this is also far away. For now there is pain—a throbbing dullness that has spread across my chest and around to my back, at the center of which something stabs me, more sharply with each breath. John is dead. He must be. Yet I do not believe it. Somehow he has escaped. Someday we will joke about how we fooled the stupid dorks; how they robbed us and left us for dead, but how we escaped to safety and recovered and finished our walk around the world.

But John is dead. That last, muffled shot can mean only one thing. "And I could have killed them!" I think. There they stood in front of me; all I had to do was pull the trigger. The image keeps coming back: me blasting them. They fall. The two in the middle do not move. The others twitch and turn, each in a different position. I break the gun, ejecting the cartridge, shove in another and click the muzzle back into place. One of them looks up, pleading for mercy. Another tries to stand and falls instead, crying out. My finger slowly presses against the smoothness of the trigger. The gun is at my waist. I take another step . . .

But John is dead. And I am lying here with a hole through my chest. Before they go someone will think, "Shoot him. Put a pistol to his head and blast his brains out." I see my brains splattered on the crumbled, reddish stone of the roadside. I cannot see my face or head or anything else except the form of my body twisted upon the ground; but I see my brains as if they were there before me, spilled out of a can.

A massive figure looms. I hold my breath, waiting for the end. He turns and goes away. I hear the sound of a truck starting up. Voices. The truck approaches. Stops. Running feet. A voice barks commands. Things being loaded. A door slams. The truck drives off, fast, down the road toward Kabul.

A trick. They have left someone. If I twitch a finger the man who stands above me, watching, will blow my head off. I must wait him out. I must lie here and wait for a sign, wait until I hear him leave. I wait. I lie and wait for hours, it seems. At each long, careful breath, the pain turns within me like a sword, twisted. Finally, I allow my right eye to crack open. No foot stands next to me. Slowly, ever so slowly, I turn my head. There, twenty feet away, is John. He lies curled up on his left side, his back to me. Moonlight bathes the palm of his upturned left hand. How loosely it lies. How natural he looks. As if asleep. A tremor passes through me. He *has* fooled them. "John!" I do not care who hears. "John!" Silence. I stumble to my knees, crawl over the rough ground to him. I touch his shoulder and he turns, too easily. There, behind his face, there is nothing— a pulp, a can of spaghetti dumped onto the ground. His face is there, but it is like a mask, for there is no head: only a mush, an oozing, a spilled, broken squash of a nothing. Automatically, I feel for his pulse. There is no beat. I press my ear against his chest. Nothing. I stagger to my feet and cry out from the pain in my chest.

I want to cry for John. But it's too late. I have lain too long in silence with the knowledge that he is dead, and now the sob will not rise within me. The pain in my own chest is too severe, anyway. To keep from fainting, I have to bite down on my lip until I taste the blood. For John it is over, at least. John is with God. "But with you it is not over," I say to myself, "so don't be stupid. You will live through this no matter what; later you can cry. What good does crying do, anyway? Keep calm and think. How can you get help?"

I decide to cross the highway, sit and lean back against the barrier, and wait for the next Afghan truck. The thought of sitting makes me weaker than ever. All I want to do in this world is get across that road and lean up against that stone wall. I can feel its coolness against my burning cheek. I stumble across the road and let myself collapse against the barrier. But leaning back makes the pain worse. I squat, my shoulder propped up against the low wall. Waiting, I think of nothing except the truck that must come in time. I will it to come now, to come soon. Finally, I see headlights down in the gorge. With the help of the barrier, I get to my feet. The headlights blind me. I can barely get my right arm above my head. The truck slows down, then, with a roar, speeds past me. Of course, I almost laugh as I collapse once again. *I* am danger. *I* am bandits! I imagine the driver's terrified face; his red-rimmed eyes search the shoulders expecting, at every turn, the roadblock.

"Why not ride out?" Slowly, I force myself to stand. I manage to cross the road to where Willie is still chained. Reaching into my pocket for the key to the padlock, my hand trembles so that the key falls to the ground. Desperately, I pat the stony earth, searching for the key. All my hopes are pinned on finding that key. The key is my salvation. Everything will be all right if I find it. But I cannot find it. It is gone forever, and I am lost.

I find the key! This time I get it into the lock, and all the while Willie, our good mule Willie, quietly stands there, neighing softly. When I try to get up on her, though, she moves and brays a warning. I have forgotten that she will not be ridden. Anyway, I cannot mount her. And if I did, where would I go? It is ninety miles to Kabul, fifteen to the nearest town.

"Walk Willie over to the road! She will stop the next truck herself." I follow the command. The next truck barrels by faster than the other one. I let Willie go to fend for herself. "Run back to Portugal," I think. "Go ahead. We will not stop you now. Go. Take care of yourself, you big mule."

Squatting next to the barrier again, I hear noises from down the road. Shadowy figures, walking close together, are moving slowly toward me. I roll myself over the barrier, drop three feet, and almost slide off the embankment and down to where the river roars forty feet below. Blinded with pain, I lie there on the edge of nothing, then slowly pull myself over to where the blocks divide, and look through.

A nomad caravan. Five men, four women, three children.

The men have rifles but their eyes are on our wagon as they pass. The children are asleep, tied on top of the camels, the women, switches in their hands, walking behind with the dogs. "They could help me," I think, as they pass. "Or they could shoot me, if I show myself."

After they have gone, I drag myself back up over the barrier. "Gasoline! Pour gasoline across the highway and when the next truck comes, toss a match into it." I struggle across the highway to the wagon, stumbling over our scattered gear. I cannot find the can of gasoline. Climbing back out of the wagon, I almost faint, and think to myself, "Get back to the road. If you faint here no one will ever find you. Faint by the road if you must faint." I collapse once again up against the barrier. Inside me, something pops. I feel a warm spread of wetness on my chest. Blood gushes from my wound. "Oh, my God, help me! Don't let me die now. Please, God!" I am trembling and crying out loud.

Noises down the road. Dim figures walking toward me. I stumble to my feet and, half bent over, yell, "Help! Help!" Let them shoot me if they want. I will bleed to death soon enough in any case. They keep coming. Another nomad caravan. I plead with them. I will pay them. For the love of God, they must help me. I cry to them for mercy, for the slightest assistance. The men finger their rifles as they pass, look from me to the wagon and back again. The women look straight ahead. "Oh, God," I think, as they disappear into the night, and call upon John to help me. I am more alone than I have ever been in my life.

I must have fainted because the next thing I know I am lying on the ground, half leaning up against the barrier. I struggle into a crouching position and place my right hand over my wound, putting pressure on it with my left arm. After a while my hand feels sticky; blood stops dripping through my fingers. I close my eyes and try to sleep, but the pain pulls me awake each time I breathe. I sleep a little between breaths, it seems. What time is it? If I can hold out till morning, surely someone will find me. Someone will come to my aid. For the first time I think of our dog, Drifter. Where is he? I try to whistle for him, but no sound comes through my lips. Did they kill him, too? I wonder.

Just before dawn I hear a car. I wait for the gun of the motor when it goes by; but instead, it slows down. I open my eyes. A Land-Rover. Six men in it. Soldiers. "Thank God!" All I can think is "Thank God!" over and over again. And then when it is just past me the Land-Rover speeds off. I shout. I get to my feet and yell after them. I take two steps and then I am on the ground again; when I look up there is the Land-Rover in front of our wagon.

Two men are standing beside John's body. One of them walks over to me.

"You are hurt?" he says in English.

"Yes," I tell him. "I am shot." He looks puzzled. He does not seem to understand why I am laughing.

"Your friend here. He is dead," he says.

"Not my friend," I say, smiling up at him. "My brother."

"Come," he says, sternly. "We must take you to Kabul."

CHAPTER 2

"You can lie down later," I tell myself. "The pain is nothing. You are safe now and in two hours you will be in Kabul." Still, every jolt of the truck digs at my wound. At each turn, it takes all my strength to keep from slipping to the floor. I cling to the seat in front of me and try not to think. But I can feel my shirt getting wet again and I know that my wound has opened. "Oh, God," I pray. . . . And then, suddenly, we screech to a halt, and there in front of us is an overturned truck, one wheel still spinning, and just in front of it, not twenty feet away, even more grisly because of the headlights, two bodies. The one nearest us has no head. The other is twisted impossibly and covered with blood. They are Afghan soldiers. The headless man is still holding his rifle. Off on the other side of the truck I can make out more bodies. There must be others down in the gorge, too, I think, as I see where the truck has hit the stone barrier. I shiver and look away. We make a U-turn and soon we are back in Sarobi. Then we are at the accident again, bringing with us police and two other trucks. "We were told the gorge was full of bandits," one of the soldiers explains. "We did not want to come unprepared."

I should feel sorry for them. After all, they came to rescue us. But all I can think is "Thank God. Thank God it's not me." After what seems hours, a man with a bandaged arm and leg is helped into the seat next to me, and we take off again.

"You're going to take me to the American Dispensary, aren't you?" I say as we get near Kabul.

"Everything will be as you wish," the policeman answers, but when we get to Kabul we pull up to the Afghan Hospital. I will not get out. The Afghan next to me is helped out, but when they try to get me to go, too, I say, "No. You must take me to the American Dispensary."

The Afghan Hospital is a butcher shop. Everybody knows that. If I went in there, I might never get out again. Dr. Moede, himself, has seen them throwing buckets of water over operating tables to prepare them for the next patient.

About ten minutes later a doctor comes out. There are two attendants with him, so I brace myself for trouble.

"Please," he says. "We want only to help you. Allow yourself to be brought in."

"No," I say again. "You must take me to the American Dispensary." I am so weak now that I can barely sit up on the seat. The doctor stands there for a moment, and then he turns and leaves. Another ten minutes go by. Then out comes a police captain. He opens up the door and leans halfway in.

"You must come with me," he says. "Later, we will take you to the American Dispensary."

I don't know where I get the strength from, but I manage to drag myself up straight enough in the seat so I can look him right in the eye. "You take me to the goddamn American Dispensary," I tell him, "or I'm going to die right here in this fuckin' truck." Well, that does it. His eyes bulge and he steps back, and then, in a very different voice he says, "Please, sir. Don't die! We will take you." They do, too. Fast. Things are bad enough for them now. They don't want to be stuck with another corpse.

When we reach the American Dispensary they let me out and drive off. They probably hope I'll die right there in the street. I stagger over to the gate, but—what do you know?—it's locked. So I hang on the bars and yell, and finally the watchman comes over and unlocks it for me.

I push the red emergency button that alerts Glenis Nielsen, the American nurse who lives down the street, and then I sink onto a seat nearby. What a sight I am! Shirt and pants, right down to my crotch, soaked in blood. But I am here. I've made it.

"You're a lucky man," says Dr. Moede, an hour later. I'm grinning like an ape. Can't stop. The bullet has passed right through me, taking a nick out of my left lung, but missing my heart and backbone by fractions of an inch. "You haven't got much blood left, but you'll make it all right now," he says. "Anything else we can do for you?"

"Could I have a Coke?" I say.

I'm feeling high as hell from whatever he's given me, but also very thirsty. He looks sort of startled, so I tell him that with a hole in my chest I hadn't been sure whether I should drink anything or not.

He laughs, "Anybody that's been through what you have, gets whatever he wants."

Later, in my room, Glenis and another nurse give me a bath. Does that ever feel good! Then Ambassador Neuman and his wife stop by, but I can't say much to them, I'm so sleepy.

When I wake up it's dark again outside. I drink glass after glass of water and then, with Glenis holding me around the waist, I cough up globs of dried blood. Each cough seems to tear my chest apart, but I have to do it, she says. So I do. But oh, I'm tired. Just getting out of bed to piss is like climbing a mountain.

Three days later I find out that Dr. Moede was more worried about me than he'd let on. My blood count had been down to practically nothing when I came in and they weren't sure how much I was still bleeding internally. I'm definitely recovering, though. Two days later, Dr. Moede sticks a big needle through my back and takes about a quart of blood out of my chest cavity. I can breathe again. By the end of the second week, everyone we know in Kabul has been in to see me, and the nurses and I are having a ball. Drifter's OK, by the way. The embassy people found him that next morning when they went out to check our stuff. Our friends the Tracys have got him now. Want to keep him, they tell me. So I guess he'll stay here in Kabul. Good old Drifter. Good dog. The walk's no place for you either, is it?

It's sometime during the second week that Witt Azoy comes in. He was the officer in charge of us for the three weeks we were in Kabul and a special friend of John's.

"What happened, Witt?" I ask him. "Why didn't they send out someone to guard us if they knew it was so dangerous?"

"I picked up everything I could find," he says, putting a big, blue plastic bag down next to the bed. Then he looks over at me. "No one's told you?" he asks. I shake my head.

"It's because you were walking for UNICEF. They thought you were carrying a lot of money. The papers. They got the story wrong. They thought you were *collecting* money for UNICEF, not just handing out pledge cards."

"God damn," I say.

"God damn is right," says Witt. "They'd been following you for three days."

Bastards. So I should have shot them after all, the minute I saw them coming up. "I don't suppose they'll ever get them," I say.

"I don't know," says Witt. "They say they're looking."

Suddenly, I see the six of them hanging there next to each other on the gallows, each twisting in his own space—black tongues jutting out, eyes rolled up, popping. A foot twitches. A hand contracts. Their bodies turn slowly this way and then that. It's because of me that John is not alive. I look up at Witt. "God damn," I say.

"You know what John told me?" says Witt. "That if he knew he was going to die the next day, he'd still keep on walking."

"Yeah?" I say. Quickly I put my head down so that Witt does not see the tears that have welled up in my eyes. He turns to leave.

"So long, old buddy," he says. I try to say something but the words will not come, and then it is too late and he is gone.

After a while, I start to go through the stuff Witt has collected from the scene of the shooting, and one of the first things I come upon is a notebook of John's. One passage in particular I read over and over again.

> Of late I've been having scary thoughts about the rest of our walk through Turkey. We've talked to many people in the last few weeks and most of them think we were crazy to go into eastern Turkey at this time of the year. There have been many warnings; the wolves, the weather, the dogs, the terrain, and even the people can defeat us. The possible dangers are a little scary, but the real scare comes from the frequency of the warnings and our blind determination to push on in spite of them. Our attitude has always been thanks for the warning but we've committed ourselves and we have to keep moving. It's this going on in the face of almost overwhelming advice against it that bothers me. It's like a play with all the actors playing their parts right up to the tragic ending.

It's hard for me to believe it's John writing this. Each country we went through there were always people telling us how awful it was going to be in the next one. But that never bothered me. I didn't think it bothered John. The Turks warned us we'd be murdered for sure crossing Iran. And the Iranians couldn't believe we were going to walk through Afghanistan. We used to laugh about such things. It was only the Afghans, in supposedly the worst country of all, that did anything to help us. They had soldiers walk with us for our protection. Of course, the only time we needed them they weren't there, but . . . I read on. I haven't read any of it before and it's kind of like talking to him. I even read some passages out loud.

"What am I doing here?" writes John. "That's a good question. Somehow or other I enjoy this stupid life, even though sometimes I hate it."

"Sometimes is right," I say. "Each time I pull you out of bed in the morning. Right?"

"Of course, when I say I was fascinated with the idea, that takes in a lot. When I first heard of it I said, 'I don't think that's ever been done before, at least I've never read about it.'"

"What do you mean?" I say to him. "Rich and I looked it up in the *Guinness Book of World Records*. In a couple of other books, too. Of course no one's ever done it. Do you think we'd be doing it if they had?"

"Somehow it's always been in me," continues John, "this desire to do something different, maybe to seek something without knowing what. Going on an 'adventure' without really being able to define what an adventure is. When I was about eight or nine years old and living in Stewartville, Minnesota, it came out, but was never fulfilled. My friend and I decided we would use the wood scraps in his backyard to build a raft and use it to explore the river that ran through the other side of town. . . ."

"You were going to go to Florida, remember?" I say. "Just bum around for a month or so. But you couldn't get anyone to go with you. You wanted the walk as much as I did, didn't you?"

"It would also have a lot to do with luck, as far as our making it goes."

Luck! It's a subject John and I argued about a million times. I close up the notebook. I don't want to read any more tonight.

"There are many reasons for my being on this walk," John had written. "But of them all, the search is the most important." He'd found it, hadn't he? Whatever it was. That's the important thing.

Two days later, I'm chatting with Paul Fleming, an English friend of ours who has come to visit me in the hospital. "I hear you're planning to finish the walk yourself," he says, not looking at me.

I nod. It has never crossed my mind to do anything else, actually, but a lot of people seem surprised when I tell them.

"Well," he says, looking like an anxious rabbit in somebody's cabbage patch. "I'd like to go with you."

"How can you, Paul? I mean, you've got a job." I can't tell him straight out there isn't a chance he could make it. I can't tell him either that it has to be my other brother, Pete, who finishes the walk with me, or no one. That's too complicated to get into and he probably wouldn't understand it. It's a family thing. Besides, Paul is a skinny little guy who looks as if a good strong wind would knock him down.

"If you don't want me to go with you, just say so," he says.

"It's not that," I say, trying my best to make my voice sound as if I mean it. "It's that another guy made me promise I'd take him. Thanks for the offer, though, really."

Well, he looks sort of half relieved when I tell him that, so it ends up OK.

A while later a couple of marine buddies come in and they've sneaked in some booze, so we all have a little party. Strictly against the rules, but hell, I'm feeling pretty good by this time. It's funny with a guy like Paul. You just know he won't make it, that's all, and there's no use pretending otherwise. It's like the donkey and cart they gave us in Portugal. We had a terrible time getting them to take them back and get us a good big mule instead. And yet anyone could see that that little donkey wouldn't be able to get across Portugal, much less around the world. Hell, we had to practically push it to Lisbon. But you couldn't tell them that. Not the mayor of Cintra, not all those people. Not when they'd been so nice and made such a fuss! But even if Paul were up to it physically, I still wouldn't want him. It has to be Pete, or no one.

My father calls the dispensary one night from the Waseca radio station. He's been trying to get through for over ten hours. I'm still kind of short of breath, but we talk for about ten minutes and I speak to Mom, too. They want me to come home. Dad has already sent the money for John's body to be flown back to Minnesota and for my plane fare, too. I tell him that I'm going to finish the walk, that now that John is dead it's all the more important that I don't quit; and after a while he says he understands and that he won't try to stop me. I tell him he'll have to promise to fly me back again if I do come home. He says he will. Poor Dad. Once again he's shelling out money he doesn't really have for something he can't fully understand the significance of. It's Mom who keeps pushing him, I know; but then, when we first told them about the walk, Mom actually wanted to go with us.

Once everything is settled, though, I feel good about going home. Two years is a long time to be away from your folks. This way I'll get a good break, be able to be with my kids over Christmas, and have a chance to talk to Pete. As long as I come back again and finish the walk!

One morning, a few days before I'm to be discharged, who walk in but Carol and Patty. They met, accidentally, in Tehran, heard about the shooting, and decided to come back and see me. Well, I can't help but be touched by that, but at the same time I'm mad at Patty for coming back after all I've done to help her get out of here. With Carol, it's different. She's come for John's sake. We haven't talked for ten minutes before I'm as miserable again as I was right after the shooting. All John had talked about those last three days was Carol. I'd seen him fall for a lot of girls in the past two years, but none the way he had for her. And here she is.

She's come all this way just to hear from me how it happened. She hopes, too, I know, that somehow it isn't really true, that John is still alive after all and that she will stay with him this time, stay with him forever. But it isn't to be. I have to force that knowledge on her, and so it all comes back: his upturned hand bathed in moonlight, his face—with nothing behind it. I have to comfort and assure her that John is with God, that *he* is happy, that it is only for the living you should cry. And all the while all I can think of is him lying there.

With Patty, though, it's different. I don't know why she's come back. It isn't for me, of that I'm sure. Maybe it's just because Carol needs someone to be with her. That could be. What I really think, though, is that she can't stand to be away from Kabul. There's something in her that loves all this filth. Patty's weird. I knew that the first time I met her. That's why I never went to bed with her. She's had too many other men, too many other ways; and drugs have knocked her too gaga for my taste.

The night we met she was sitting on a barstool in the Chinchilla Club. She was wearing one of those long, flowing Afghan gowns. We'd been in Kabul only a little over a week, and we hadn't met any girls yet. She was pretty, in a thin, wispy sort of way, and she had great eyes: very large and deep and almost black—sad and soulful, but sort of wise-looking, too. She'd left home—Hempstead, Long Island—when she was seventeen and had been traveling around ever since. She talked for a long time about how rotten her parents were and how everything in America was so plastic and nothing, and how nobody she knew got any fun out of life. And then, abruptly, she left. She had to check out the 25 Hour Club, she said.

I finished a couple of drinks, thought about her, and decided she wasn't my type. Then I got up and followed her over there. I was horny. But nothing ever came of that. Instead, I became sort of her father confessor.

"What are you going to do?" I say to her. "Go back to Ahmed? Those three Indians you told me about? You going to look them up?"

"David. Please." She's on the verge of tears. That's one of the troubles with her. You aren't supposed to act any way but just very understanding with her, no matter what.

I don't know where she spent the next few days, but the day I check out of the dispensary I bump into her at the 25 Hour Club. She's pretty drunk, and she comes weaving over to me and sits on my lap and sort of curls herself all around me. We make out a little, but when she pushes her hair back from her face I can feel her cheek is wet against my neck.

"What's the matter, baby. You having a bad time?" And then she tells me. She's been staying with Maureen, a friend of hers, but all Maureen is interested in is her next fix, and the Afghan she's getting it from has started to put the make on Patty, too.

"I'm scared of him, David. Really scared, you know. He could rape me, I know he could. You remember Sarah—Sarah Bishop? Well, they found her in the gutter last week. Dead!" Patty is in some sort of trouble with an Afghan, I know that. From the way she looks it almost certainly involves drugs. She doesn't fool me. It's probably Shamboo. Patty's been raped by Afghans several times. She's probably forgotten half the stories she's told me. But what the hell. I'll be staying in the Kabul Inn for a week before I leave for America, so I say, "Do you want to move in with me?"

She acts as if I've saved her life.

I think it's good for me to have Patty around, because worrying about her getting out of Afghanistan again takes my mind off my troubles. As before, it isn't a sexual relationship. I'm not exactly in shape for that anyway. But it's nice to sleep with a woman again; to have one in bed, and to talk to. On November 10, five days after my discharge, I put Patty on the plane to Herat. From there she can catch a bus to Tehran, where I'll meet up with her. If all goes well, we'll travel to Istanbul together on the train. I don't feel like just flying straight back to America.

Before I can leave, though, I have to talk to the Chinese embassy. When I told Mr. Hoelgard, the UNICEF man in Kabul, that I planned to continue the walk with my other brother, Pete, he had tried the Chinese embassy again, and this time they were interested.

To walk across their country? Where would we enter? Where would we leave? How long would it take? What would we need? I mention walking with two Chinese brothers. That had been one of John's ideas, and Mr. Hoelgard thought it was great. So do the Chinese. After all, Ping-Pong has opened the doors, right? Whether it's that or the shooting, or the fact that the Russians have denied us permission to cross their country, I don't know, but they are very polite and cordial, and they say they will ask Peking and let us know. It looks pretty good and we're pretty excited about it. It would be great publicity for the walk as well as for UNICEF. Pete would like the idea and it might make his wife a little easier to handle.

On November 13, 1972, I fly to Tehran. But Patty isn't there. I wait for her for a few days and then give her up and fly straight to America.

The first half of the walk is over. For John, his earthly life is completed, but I find something in his notes that makes me feel better about that. About the walk he'd written: "It was an opportunity to rise above the normal trivialities of life and do something unique. One of those chance-of-a-lifetime things." So John and I felt the same way about that, at least. To die doing what you wanted to do was a hell of a lot better than to live in the dull rut of everyday routine, the way most people did. That was for sure.

CHAPTER 3

When I get off the plane at the Minneapolis Airport, there are so many TV cameras, radio mikes, and reporters, I look behind me to see who the big shot is. It takes me a while to realize the big shot is me. We've gotten attention before, but nothing like this. My family is there too, of course, and what I really want to do is see them, but I have to hold a press conference. Right away! Imagine! First of all they set up this super-romantic picture of my reunion with Jan.

"Would you make that smile a little bigger, Mr. Kunst? That's right. Just a . . ." Click. The big movie star kiss. Click. Click. Then they line us all up behind a desk and make me go through the whole shooting thing again. "And how did you feel, David, when your brother was shot?" Well, long before, I'd made up my mind to play the game of returning husband, but I had no idea it would be this bad.

The kids are great, though. They want to hear about everything. Danny is still too young to ask many questions, so he just sits on my lap the whole way back to Waseca; but Bradley and Debra never stop.

"Is Princess Grace really as pretty as you said in your postcard?" asks Debra, and I tell her "Prettier." Bradley wants to know all about Buzkashi, the national sport of Afghanistan. The bloodiest game ever invented. He's only eight and all teeth, just like Danny, but already he's hung up on violence.

"It's kind of like football," I tell him, "except you're on horseback."

"And you use a dead goat?"

"Oooh," says Debra.

"Or a dead calf." I don't tell him that it's usually beheaded.

"And they just pick it up and drag it back and forth?"

For a minute I am there—the dust, the crowds of milling Afghans, the desert heat. Two riders close in on a third who has the carcass slung across the horse in front of him. The goat is knocked to the ground, and in trying to recover it the rider falls, his foot catches, and before his horse can be stopped, the man's head has been battered to a pulp on the rocky terrain. One of the other riders gets the goat's carcass, however. Amid wild cheering he carries it back to his side and makes a circle. A goal for them. Later, we are almost run down when a group of riders plunges into the crowd next to us. A boy and an old man are killed. The game goes on, though. There is not even a pause in the play.

My mother tells me about the funeral. Seven priests. All friends of the family. Hubert Humphrey was there. She wants to know if I would like to stop at the cemetery on the way home.

"No thanks," I say, and her mouth tightens up a little in disappointment. I'm glad to hear they have put "World Walker" on the gravestone, though. John would have liked that. Mom's put on weight since I left. Dad hasn't changed. Debra's head is nestled under my arm. While my mother talks, I study the smooth part in her shiny brown, straight hair. All the kids have grown about six inches.

Mom and Dad leave for Clear Lake right after dinner.

I'd hoped they'd stay the night. So after the kids are in bed, I'm left alone with Jan.

"What's this?" I say, picking up a scrapbook that's lying on the table. The whole thing's about the walk.

"Oh, I made it," says Jan. "After work. In the evenings. When I had spare time." She brushes the hair from her forehead with the back of her right hand, sighs, and gets up. "More coffee?" she says, but I don't answer. There they all are, in order, the date neatly entered at the top of each page. There's the marquee with "Site for the Start of the World Walk" pasted up on it in two-foot letters. There's me shaking hands with Hubert Humphrey, John standing by, grinning. Meeting Princess Grace of Monaco. I laugh out loud when I get to us and Willie Make It inside that restaurant in Italy with Thor Heyerdahl. And there we are in Venice, where no mule's hoof has ever trod. Most of these clippings I haven't seen, and I read through the scrapbook from start to finish, hypnotized. I am vaguely aware of Jan coming back in, of her going upstairs, coming down again, of her sitting in her armchair on the other side of the room, knitting. "Hey, look at this!" I'll say, and maybe she'll get up and look down over my shoulder at whatever it is, and then go back to her seat again; or maybe she'll just say, "What's it of?" and keep on with her knitting.

Toward the end there are pictures of the funeral. "We hold no hate in our hearts," my father is quoted as having said. "We believe in what the brothers were doing, working for world peace and understanding through UNICEF. We hope that somehow through John's death and Dave's injury their mission and UNICEF's mission will be furthered and that during this Halloween season people will remember John and Dave and what they were trying to accomplish by walking around the world for UNICEF."

"Now why did he say that?" I say. Jan looks up, puzzled. I read the passage out loud. Had he really said that? I wonder. What went through his mind when they told him? Thoughtful. Squirrel look. Head a little to one side.

"What do you mean?" says Jan.

"It's all wrong, that's all. We didn't walk for UNICEF. That idea came later."

"Then why did your father say . . ."

"I don't know. That's just it. Maybe someone just . . ." But Jan isn't listening. Her face is set.

I go out to the kitchen and get myself a beer. "So, how's your job?" I say. "The new boss OK?"

"He's nice," she says. "We all got a raise."

I sit down lightly on the arm of her chair and put my arm around her.

"Don't," she says. "You'll break it."

"Want to go to bed?" I say, taking a swig of beer.

"Uh, huh." She doesn't look up.

"Why am I pretending?" I ask myself. I don't really want her.

She yawns, puts her knitting down, rolls it up, and places it back in the basket next to her chair. "Turn out the lights?" She puts the scrapbook I've left lying on the sofa back on the coffee table, takes her coffee cup to the kitchen, and holds out her hand for my beer bottle. I give it to her, and try to catch her eye as we walk together around the room, but she never looks up. I could say, "Hey, Jan," and make her look up, maybe, but then what? Probably she'd just say, "What do you want?" I follow her up the stairs and notice how much thinner her waist is.

Jan falls asleep right away, but I lie there, staring at the ceiling where the streetlight makes an oblong out of the square of the window. Every once in a while I can hear a truck coming through town. I lie there thinking of the walk.

The image of Michele forms itself in my mind.

Laughing at my objections, she runs down the beach—nimbly dodging towels and children with their pails—and plunges into the water. I follow her, but more sedately.

Standing waist-deep she turns toward me, flaunting her bare breasts. I must be showing my disapproval, for she runs toward me, splashing water. I dive and take her by the knees and drag her down, and when we are further out we embrace. Her nipples are hard as peas against my chest. Even here, and though the water is cold, she stirs my sex. "You Americans are such prudes," she nuzzles her whisper into my ear. We swim way out past the life raft, past the crowd of bathers, until a whistle blows and we are beckoned in. Nice, the playground of the rich. We have been staying with Michele and Janice, her sister-in-law, for the past three days —in the garage of friends of theirs from Paris, where there are two cabin cruisers stored.

John and Janice have taken the tent. It is our first night alone! I bolt the door and note that the only window is barred. When I get into the cabin, Michele is already under the covers. I take off my clothes in the dark and slip in beside her. Slowly, we explore each other. Her shoulder blades are tiny wings. I nuzzle her breastbone, feel the high arch of her instep with my big toe. Inside her, I long to burst. We start our race close to its end, and finish together. The urgency gone, time is infinite—we soar, we melt, we burrow, we devour. The night slips by like soft wind.

Suddenly, loud banging on the door, voices. "My husband," whispers Michele, her voice trembling. "Oh, my God," I think.

"Into the other boat, quick," she says. While I crouch in the cabin, an empty whiskey bottle in one hand, a fishing spear in the other, Michele tries to explain things to her husband. The longer I wait there the more ridiculous I feel. Surely, any second now, he will come rushing in. "Where is he? Where is the pig? I'll tear him into little pieces." But nothing of the sort happens. I go from cold sweats to hot flashes and then, all at once, Michele is at my side, murmuring "It's all right." The husband is appeased. He and his friends have gone off; they will not be back until morning. Oh, wonderful Michele. *Ma belle.* How do you do it?

"Of course, I'll leave early. I'll leave now. You're sure he won't be back? You're sure?"

"Will you please stop moving." Jan's voice, grumpy with sleep.

"My wound. It still hurts me sometimes," I tell her. She grunts and turns back on her side. I try to lie still, think about something else. Eventually I fall asleep, only to have a terrifying nightmare.

John and I have just crossed over into Bulgaria. A man comes up to me in the street, looks over both his shoulders to make sure there are no police around, then hands me a large jar of yogurt. I thank him, but I have forgotten that nodding your head up and down means "no" in that country, so he goes off with a puzzled look on his face. When I realize what has happened, I want to run after him and explain, but he is too far away by then and, in any case, I cannot leave Willie. John, for some reason, is not there. Then I hear a shot, and two policemen come around a corner with the man who gave me the yogurt held between them. As they come nearer I see that the man has no left eye. Blood trickles down his cheek. When they come up to me, one of the policemen says, "Did this man give you a jar of yogurt?" I am about to say "Yes," when the man winks at me with his right eye, so I change it to "No." The policeman puts his pistol to the man's head and blows his brains out.

I wake up in a cold sweat, thinking that I've been screaming, but Jan has not stirred. I get up and go to the bathroom more to get the terrifying dream out of my head than anything else. Moving around the silent house, I secretly hope that I will waken one of the children. The house has a funny smell, it seems to me, a stuffiness to it I don't remember; cabbage boiling on the stove, perhaps; the slight rankness of a clothes hamper. I laugh at myself for noticing this, for being so finicky. Back in the bedroom, though, I crack open the storm window, enjoying the sharp, cold slit of air that knifes into my stomach. Not much snow on the ground for this time of year, I think, as I pad back to the bed.

When I get up in the morning the house is still. There is a note from Jan under the sugar bowl on the kitchen table. "David. Danny gets back from school *at two o'clock. Be sure to be here.* Jan." I make myself a big breakfast: scrambled eggs, bacon, home fries, four slices of toast; and after eating it wander around with my coffee cup, looking at things. There is new wallpaper in the bathroom—chubby little pink babies with lambs and ducks and ponies. The green sofa Uncle Pete and Aunt Ruth gave us for a wedding present has new armrests on it. There're another couple of bowling trophies on the bookcase, it looks like. Bowling! We couldn't even do that right together; and she won most of the time, too.

I fill up my coffee cup, sit down at the kitchen table, and wonder what to do with myself until two o'clock. Go visit Willie! No car. Jan's taken it to the shop for something. Never mind. I can walk. It's not more than ten miles. Once outside, walking, I suddenly feel happy. It's a nice, brisk winter's day, around sixteen degrees. No wind. It feels good to stretch my legs again. The walk takes me two hours and twenty minutes.

Willie—Willie Make It the First, that is—is the mule that got John and me all the way to New York City that first leg of the trip. I haven't seen her since we brought her back here in the U-Haul-It, two years ago, almost to the day. She's on more or less permanent loan to the Miller family, who live on a big farm outside of Waseca.

Mr. Miller is home, and Damien, the youngest Miller, is delighted to give me a demonstration of how well he can ride. Willie's really more donkey than mule. I am surprised how small she is. She looks great—fat and sassy—and seems to remember me. I lead her around a little for old times' sake and then we go back in the house and Mr. Miller and I talk quietly about crops, about his family—twenty-two kids they have—about Waseca—there's a new Oldsmobile distributor, Carl Swanson's still mayor. All the time, though, my mind's on Willie and the walk to New York and how hard Willie was to train. I think of the mule-naming contest, of all the preparations, of the day we finally got off.

Owatonna. Sixteen miles. We set up our tent in the town park and were too bushed and sore to do anything but crawl into it and fall asleep. Three weeks and our blisters will turn into calluses, six weeks and our legs will stop aching. That's what the mailmen said. Five weeks and six days from now we'd be OK.

It rained during the night so both our sleeping bags were soggy. I pulled John out of his at six o'clock and we got going at ten to seven, me leading Willie. We had our first fight. John said I led yesterday and today it was his turn, and I told him we'd never get there if he led and we'd better get that straightened out right now. I packed Willie, I got up first, and I led. I grabbed the rope and took off, and after a minute John ran up and I got ready for trouble, but all he said was "OK, you stubborn bastard, do it then." I didn't even turn around, so he flicked me in the ass with the switch he had for Willie; but I just picked up the pace a little. In a while he started to sing "Nobody knows and nobody cares" in his dumb, singsong monotone and we were both waving at the traffic like crazy and I joined in and about eleven we got into Clairmont and a guy came out of a restaurant and invited us in to eat. "Quite a crowd they had for ya," he said, meaning the send-off we got the day before in Waseca. He'd seen it on TV. There was a picture of us in the morning paper, too.

"Yup," I said, taking another forkful of pancake.

"Hubert Humphrey was supposed to come, but at the last minute he couldn't make it," said John. "He's a pretty nice guy. Said just to call on him if we needed any help." Boy, you could see the man's eyes pop out of his head at that. His wife was over at our table now, too, if that was his wife. And there were some locals milling around in the background, straining their ears.

"We been readin' 'bout you in the *Minneapolis Tribune*," she said, nodding a couple of times. "Robert T. Smith. He sure thinks a lot of you boys."

"He's pretty special himself, ma'am," I chimed in. "If it weren't for him, we'd have walked out of town and nobody'd've even noticed we'd left."

The man laughed and looked at the lady, and you could hear some light snickering from the back rows. Then a man's voice yelled out, "Two asses and a mule. That's what I hear they called you." There were a good many guffaws at that. . . .

"The asses I know are glued to the barstools," I said, which got a pretty decent reaction.

"Don't you listen to them, young man," said the lady. "What you're doing makes us all real proud." She was kind of fat and dumpy and about sixty, but nice. You could tell she was worked up about it all.

"'Willie Make It.' Pretty funny name for a mule," the man said, filling up our coffee cups.

"Will she?" said one of the locals.

"Looks pretty small to be walking around the world," said another.

"Will you, is what I want to know," said an old geezer. "A mule's stubborn."

"One step at a time," said John. "Just one step at a time for all of us."

It was a good feeling. We were in that restaurant about an hour, I guess. The old lady made us take some cookies that she'd baked herself, and they all came out to see Willie, and when we walked off there were about fifty of them waving good-bye to us and wishing us luck, and some of them were cheering. We loved every minute of it, I have to admit. After months of no one taking us seriously, we were celebrities now, no matter what happened. People had been stopping to talk to us, or just to wish us luck, all along the road. If it kept on this way, we'd skate over those thousand miles to New York as if we were on ice, no matter how much our feet hurt. The tough part would come later. But we'd worry about that later, too.

We crossed over to the right side of the street, and then instead of going east, Willie just stopped. I pulled at her, but damn it, she wouldn't budge. John hit her with his switch, but she didn't even blink. So I turned her in a circle, thinking that might get her going. No good. She plain refused to go east. When the crowd saw what was happening, did they ever think it was funny. To make matters worse, Pat O'Leary, from the *Waseca Journal,* drove up and started taking pictures. I led Willie back across the street to where the crowd was, thinking that all the noise and commotion might cause her to get a move on, but when I tried to turn her, she just stood there. Well, I wasn't going to let the walk come to an end there because of a stupid mule, so I yelled out, "Probably she's jealous because she didn't get any ice cream."

"Probably she figures there're better things in the other direction," someone shouted back.

But in a few seconds the man who ran the restaurant came out with a big bowl of vanilla ice cream, and I said, "Just let her get a taste of it and then kind of lead her on with it," and he did, and it worked. We just kept right on walking on the wrong side of the street till we got way out of town. Then we crossed over and everything was all right. Maybe that's all it was. Maybe Willie *was* jealous. You can't ever tell about a mule, that's for sure. At least it made a funny story and that helped us, we found out.

Any publicity is good publicity. We were quick to discover that. Particularly if we wanted to get across the fact that offers of room and board and anything else that might come to people's minds were most welcome. We wouldn't get as far as New York, not to mention around the world, unless a lot of people helped us out.

It began to rain as we came into Rochester, Minnesota, and we were feeling pretty glum, what with it being almost night and no one there to greet us, when Jess Little and his wife drove up and invited us to spend the night. Jess had been a buddy of mine in high school. I don't know why I hadn't thought of him before. Boy, did it feel good to take a shower and get rid of Willie and sit down to a big dinner and *rest our feet*. We'd made twenty-two miles that day and John was limping pretty badly and my feet were sore, too, but what with plaguing John the way I had I couldn't very well complain myself, so I just pretended that mine felt fine. I was getting ready to nod off after dinner when suddenly Jess got really excited. We'd been talking about the trip and Jess was pacing up and down the room, interrupting us and asking questions, when suddenly he turned to June and said, "By God, I'm going to do it!"

"Do what?" his wife said, kind of nasty.

"Take that canoe trip down the Mississippi. I've been talking about it long enough."

"Well, you just go right ahead and go," she said, and stalked out to the kitchen.

"I might just not come back in a hurry, either," he said, loud enough for her to hear him even if she'd shut the door.

She came storming back, then. "And if you think for one minute I'll be sitting around here waiting for you, you're crazy. Some wives are different," she said, looking over at me.

"Now, wait a minute," Jess shouted, but not as loud as the last time. "I didn't say . . ."

"You didn't say anything. I know. And you better not either. And there're a lot of people, David Kunst, who don't understand how you can walk out on your family like you have and think it'll all be just fine." She glared at me but I glared right back, and then she turned and marched upstairs.

"Women," Jess said, and shook his head, but he didn't fool us any and he knew it; half an hour later he followed her up to bed, and I bet it was an hour before they quit yakking at each other.

"You know what Jan said to me when I told her?" I said to John. We were in our sleeping bags on the floor.

"What?"

" 'I always knew you'd do something crazy like this some day.' That's what she said. She didn't even try to talk me out of it. After eleven years she knew me better than that."

"So what did you say?"

"I asked her if she wanted to come along."

"You didn't!"

" 'Not on your life,' she said, just the way I knew she would.

" 'What'll you do, then?' I asked her.

" 'Go back to work.'

"It was easy, I just wish I'd done it sooner."

"You were lucky," said John. "Judy probably still thinks I'm coming back to her when it's all over."

"You're not married," I said. "That's different."

"She thinks we are. We've been living together for almost two years."

"Well, just remember, you don't owe her a goddamn thing."

John yawned. "Big brother, I thank you for all your advice and comfort and leadership, and let's get the hell to sleep."

"OK. Forget it." I rolled over onto my side. After last night, the rug felt soft, believe me. I think I was asleep in two minutes.

CHAPTER

4

RIGHT THROUGH CHRISTMAS AND INTO JANUARY EVERYTHING is OK. I'm getting stronger all the time and it's vacation so the kids are home. Maybe three times a week I'll give a talk or a TV interview or something, and weekends we'll drive down to Clear Lake to be with my folks. But around the middle of January I begin to get itchy to get going again. I know I can't yet. The lung specialist at the Mayo Clinic said March first at the earliest, but that doesn't stop me from wanting to. So at the end of the month I fly out to Los Angeles to see Pete. We have to discuss how we're going to finish the walk, what we can do to excite the media's interest in us again, and a million other things. He's coming all right. There was never really any question of his willingness. I just hadn't made up my mind before to ask him. What decided me was that there was a good chance, apparently, that China would turn us down. If that happened I'd have to walk through India, and I knew I couldn't do that by myself. "The land of the teeming masses," someone had called it, and I'd walked far enough in those countries to know what that meant. But what's it going to be like walking with Pete, I wonder? I'm not at all sure.

August 26, 1970. We were camped beside the road just the other side of Everett in western Pennsylvania. A car slowed down. I must have heard it in my sleep, because I was out of the tent by the time it stopped. A few seconds later I heard the door slam and then I was blinking from the lights and thinking fast. Steps. Voices. I tightened my grip on the stick in my hands. I'd give them two more paces. "What do you want?" I yelled out. "Turn off the goddamn light and say something."

A pause. A familiar laugh. "Dave! It's me, Pete!" Son of a bitch! He'd hitched from Iowa to walk with us for a couple of days. John stuck his head out of the tent and joined us, finally, after I'd gotten the fire going again. Pete cracked a bottle of Scotch he brought with him, and we all sat around and had a few. It was the first time the three of us had been together like this since before I was married.

"Not since Florida," said John.

"Old Man Waters and his gators," said Pete.

"Gator got'cher," John said, and grabbed Pete's toe.

"We never did see a gator, did we? Not the whole time we were there," I said.

"Plenty of moccasins," said Pete. "I remember almost stepping on one of them."

"Come on," I said. "Not at night. They keep out of your way."

"How about coral snakes?" said John.

"You mean king snakes," I said.

"Either way. How would *you* like to find one in your bed?" So much from Pete.

"There're snakes around here, too," I said. "How do you know there isn't one crawling right up behind you now?"

"Hrrrrrrrrr," said John, trying to make his tongue rattle.

"Not rattlesnakes, dummy. Copperheads. Didn't you see that big one those Boy Scouts caught?"

"Got'cher." Pete grabbed John by the leg, and whether it was the talk or the lateness or the whiskey or what, John let out a real howl and then jumped on him. They wrestled around for a while and then I busted in and let 'em know we'd better all get to bed if we wanted to get walking early. It was about two when we turned in. What a good surprise to have Pete join us. So his wife was pissed off at him. What else was new? It had been six years since I talked to Pete, except on the telephone. Goddamn women. All they have to know is that you want to do something for them to be against it. Even brothers. It doesn't matter.

We spent most of the next morning swimming in a stream that the road'd been following for the past ten miles. Rather, John and Pete did. Someone had to stay around and watch Willie and our stuff, and I'm not much on swimming anyway. It was beautiful country there, though, and I was happy to just sit for a while and look out over the valley—watching the hawks circling in the updrafts. I climbed back up the mountain some, to get away from the noise of the road, and it was John who called *me* for a change when they were finished fooling around and decided they wanted to get off. It's funny. I can sit for an hour, two hours, as still as if I were a rock, not so much as blinking an eye, if I want to, just looking out over things. If it's nice, like it was there, and I'm alone and nothing's bugging me, my mind sort of drifts off. I don't even know it's happening a lot of the time. Just afterward, when something pulls me back—a cramp maybe, or John's voice calling me like that time. It's as if I could fly. I've been over to the other side of that valley. I've skimmed those trees, felt them brush against the tops of my toes. I circle slowly with the hawk—warm air pushing up under my belly, arms stretched out, hearing the thin whistle the wind makes through my fingers. When the hawk dives, so do I. I follow the rabbit, hear its shriek, watch its startled eyes go blank. Or I am the rabbit, scampering over the field, the dark shadow hissing above me. Flailing of wings. Pain. Blood. Wild heart beating, beating. I love views. I love views like that one, across a valley, water shimmering silver in and out of the green below, the soft line of a mountain ridge across the way, eye level; a higher peak or two beyond. No towns or anything. And the sky, a nice clear sky with a few high wind clouds. I could sit in a place like that forever.

"Hey, you guys, slow down, will you?" Pete was about twenty yards behind us. He was an ex-marine, been in 'Nam and everything, but he hadn't walked in a while, I guessed. One thing it made me realize was how our pace had picked up since we started.

"Hup, two, three, four." I sped up just a little, but I couldn't get Willie to push any faster. We were going up a pretty steep grade, and every once in a while she slipped and came crashing down on her knees. We'd have to get her shod again soon. Pain in the ass. Hard as hell to find a blacksmith that wasn't scared of mules.

"We'll take ten at the top. Pass the word."

"I'm taking fifteen here," yelled out Pete.

"Better let the recruits rest, Captain," said John. "They're mighty raw, you know."

"Now hear this, men. Shout it out, Lieutenant: left, left, left, right, left."

Pete had sat down, though, so it was no good. He has this crazy toe that acts up on him at times. The knuckle cracks. You can actually hear it when you're walking next to him. It doesn't hurt him, but he thinks it does. He thinks it will, I should say, if he keeps on walking. So he takes his shoe off and massages it. That was what he was doing now.

Funny, we all have weird feet. Mine are size 13 AA—toes like Popsicle sticks. There's one place they rub wrong, too. *Always* get a blister there. John's feet are big, also. Size 12 B. The funny thing about his is that his heels stick way out. When he played football in high school they had to pad his shoes just right or he couldn't even walk. If he didn't have pads on now, he'd be crawling down the road. I guess I can sum it all up by saying none of us would get far without Band-Aids.

John was walking back toward Pete, now, so I yelled out, "Companeee! Halt! One, two," and rounded Willie back down along the gravel shoulder. It was about six o'clock anyway, so we were looking out for a good place to stop for the night. According to the map, there was a roadside park just up ahead. I'd been hoping to get farther than fourteen miles today, but I'd settle for that if I had to. Rough walking in these mountains—harder going down than up, it seemed.

Later on, in the rest area, Pete told us he really came out to see for himself if we were serious about walking around the world. We were, he figured, and he'd give just about anything to go with us, but . . .

"Yeah," I said, "I know." It was different for Pete. We all knew that. That was one reason why I didn't ask him in the first place.

"You still could do it," said John. "Just quit your job and come. You don't like it much anyway, do you?"

"Like it? I hate it. It's nothing. I'd give anything to quit it, but what can I do? What the hell can I do, anyway? I didn't go to college the way you did. Shit." I didn't know this Pete. He'd always been mister nice guy. "Why aren't you like your brother?" I could hear my mother saying it now. Pete the peacemaker. The guy with the mop in one hand and the fix-it wrench in the other.

"Nancy'd never let you. Why talk about it?" I said.

"It's not like that," said Pete. "It's . . ."

"Sure it is. What about your kids?"

Pete looked up at me as if to say, "Having kids didn't stop you."

Pete walked with us for three days and then hitched back to Clear Lake, where Nancy and his kids were still visiting with our folks. When he left he got all emotional.

"If anything happens to either one of you, I want you to know I'll take his place," he said. He was all puffed up and serious, too, as if he were swearing allegiance to the flag or something. I kept wanting to tell him, "It's OK. I wish you could come, too." It'd be fun to have the three of us do the walk together.

* * *

That's how much I knew then. Three would have been disaster. "I'll be walking for John," Pete had said when I called him in Los Angeles.

"OK," I said. "But you don't have to."

"I want to," said Pete. "I wanted to from the beginning. Didn't you know that?"

By the middle of February things are really starting to hum. Pete and I have met with Mayor Yorty and there's been a lot of coverage in the Los Angeles papers. Letters and requests to speak are beginning to pour in again. Every day, at least a couple of contacts from the media. I've gotten hold of the Minnesota Historical Society, and not only do they make copies of all our letters of introduction as well as the newspaper clippings we've accumulated, but they give us the idea that will partially fund the rest of the walk. Serene Hanson, our biggest pusher at the Minneapolis UNICEF office, thinks it's marvelous: Friends of the Kunst Brothers. For $500 she becomes our first "friend." In all we collect $2,500—just what John and I started out with almost three years before. Not enough, of course . . .

On February 20 Pete flies in from California, and the next day we're taken by Rod Searl, our local state representative, to meet the governor of Minnesota, Wendell R. Anderson, at the capital in St. Paul. He really surprises us by presenting us with a proclamation declaring February 21 "Kunst Brothers' Day in Minnesota." He also gives us a plaque on which is written:

THEY HAVE AWAKENED THE SPIRIT OF ADVENTURE AND
GOOD WILL TOWARD MEN IN PEOPLE ALL OVER THE WORLD

On March 1, the day before we leave for Afghanistan, we meet with Hubert Humphrey. It's strange. As he's talking I get the clear sense that John is there with us; that he's standing off to one side grinning while Humphrey smiles and laughs that laugh of his, and pumps our hands.

During that last week in Waseca, there's a lot of publicity in the paper and on the radio about Pete and me taking up the walk again. What Al Austin says about us on WCCO, Minneapolis, is about as good a send-off as anyone could expect. Plenty of times later on I think back over these words:

> John's death at age twenty-five is a waste. But how many people experience as much in their lives as he did in the past two and a half years? How many will, as David wrote, "walk through vast, empty, hostile stretches . . . no water and little food, no relief from the hundred-and-thirty-degree sun and little or no security from nomadic bands . . ."? The brothers undertook this extraordinary trip for themselves first, the world's children second . . . Understand them or not, the rest of us—we sensible folks—owe the Kunst brothers something, too, and the Francis Chichesters and Walt Pedersons and Thor Heyerdahls . . . the Huck Finns without whom life would lose its imaginings.

The general feeling of the people in Waseca, though, is probably closer to what is said by a Mrs. James Sybilrud, interviewed while doing her laundry in a Waseca laundromat. "I don't like it. He's leaving his wife and three young children alone, while he takes off on some crazy scheme. If my husband tried something like that we'd be through, and especially after what happened to his brother. I don't know how he can do it." Thank you, Mrs. Sybilrud. It's you I'm leaving behind. It's you I'm walking away from.

The day we leave Waseca there's a little send-off in front of the State Theater, the same spot John and I started from. It isn't anything as big as the original, but Mayor Swanson is there and all my folks and about fifty other people. Afterward, we're escorted by a small motorcade to the Minneapolis Airport, where we have a press conference before boarding a plane for New York.

The press wants to know what countries we'll be going

through, and I tell them we're going to walk from the spot where John was killed in the Kabul Gorge into Pakistan and then, hopefully, across China, possibly with two Chinese brothers.

"How do you feel about going as your brother's replacement?" says one reporter to Pete.

"It's hard," Pete says. "I can't say it's my walk like David can say it's his. I've got to get some miles under my belt." And then he adds, "But two Kunst brothers started the walk, and two Kunst brothers will finish it."

Boy, that almost makes me bawl.

At the end of the press conference, we're presented with a World Citizenship Declaration and Planetary Passports by Mr. Lynn Elling, chairman of the World Citizenship Committee of the United Nations Association of Minnesota. John would have probably laughed out loud if he'd been there. But Pete doesn't show a thing and I keep myself under control, and finally all the good-byes are over and we're on the plane.

"So how did you persuade Nancy not to come?" I say after a while.

"Easy," Pete grins. "The money."

"How is it? She still mad?"

"Not really. She gave me a year. She's trying to be good about it. She admits she doesn't understand."

"That's more than Jan does," I say. "You know she still thinks after all this is over I'm going to come back to her and pick up my old job again on the survey crew."

"What's the matter? Don't you ever talk at all?"

"Nor do anything else together, either. Want a drink?"

Debra's the hardest one to leave, I think, as I sip my Scotch and water and look out the window at the checkerboard fields way down there that we're flying over so fast, so easily. The boys are OK. They were mostly just excited by all the fuss. But Debra sensed something. I think she knew that I'd never be back home again. Not really. Not to stay.

For her sake I wish I wasn't going off now to finish the walk, except that there was never any question of that. I didn't even have a choice. Not in my own mind. The walk was something I was going to do. It was laid out before me like the road itself, laid out for me to walk down, just as if that was why it was built. And that feeling wasn't just a part of me anymore. It was me. I could no more have given up the walk than I could have given up breathing.

Pete goes off to sleep after a while, and as I stare out at the flat countryside passing so quickly beneath us, I think of Chicago Heights, of the warning of danger ahead. Of course there was danger ahead. I smile at the memory. There was danger all around.

Some days I didn't know which I hated more: Willie Make It or the dumb, rubbernecking drivers who almost finished John and me off a couple of hundred times a day. Already one guy had sideswiped the pack. The walk was routine by this time, and mostly we didn't think about it, but every once in a while something happened to remind us that there were only about four inches between us and 200 horsepower. Of course, we could have walked on the shoulder or in the ditch, but we found out the first day that it was a lot easier walking if we were on the hard-top. And on Route 38 there were only about three feet of macadam to the right of the white line; sometimes there wasn't any. It was like walking on a wall—a curving, endless, frying pan of a potholed wall, death on one side, destruction on the other. It wouldn't have been so bad if it were just the two of us, but it was Willie Make It we had to keep on the track. On the shoulder she picked up stones, which meant she'd go lame if we didn't watch it. It was the other way she tended to veer, though. That's why I led her on my right—with my mule slobber arm—because that gave me a little more leverage if I had to come down on her, and plenty of warning if she were feeling uppity. John walked behind on the traffic side of her tail.

We'd just picked up Route 30 in Aurora when a fellow stopped and asked us why the hell we were walking for UNICEF. There was a lot of honking, and we had to get off practically into the ditch because he didn't even pull off the road all the way. So right away we were mad.

"Why the hell not?" I yelled back at him. "What's wrong with kids?"

"Commies, that's what. You want 'em to take the food outa your kids' mouths?"

"Commie kids got to live, too, you know," piped up John.

So the guy put his head back in the window and spat out, "I see you fellers been brainwashed," and drove off, tires spinning, practically causing a head-on with a pickup truck that was passing him. We thought that was the end of him and good riddance, when we saw him U-turn right next to us and there we were all over again.

"Ain't goin' through the Heights, are ya?" he yelled out.

"Yup," I said, hardly looking at him.

"Wouldn't do that, feller. Ain't hardly safe to drive through there, never mind walking."

"Oh? Why's that?" said John.

"The niggers, boy. The niggers. Don't you know nothin'?"

He went spinning off again and I yelled back at John, "Jumpin' Jesus, watch out for the jigaboos!"

"Boy!" belted out John, and on we walked.

Then the next day it happened again, only it was several people who stopped to warn us, and they weren't all crazy like the first guy.

"You think we should give it a miss?" said John that night as we were lying on the beds watching TV in the air-conditioned splendor of the Valley View Motel, where we'd wangled a free room.

"There's no other way, unless we go right through Chicago," I said. "You want to waste a whole other day?"

"No. I guess you're right," said John.

Neither of us brought up the subject again, but the next afternoon as the area got more and more run-down looking and the blacks started taking over on the streets, I felt myself getting ready for something. Suddenly, it came. A kid ran out into the street right in front of Willie and hit her on the neck with a pair of shorts he was holding in his right hand.

"Hey, watch it!" I yelled, but he was gone around the corner. There were some other kids standing on the sidewalk who laughed. In the next block there were some hood types standing between two cars on our right. As we passed, one of them said, "Whitey got him a *mule*! What you think about *that*!" Was the laughter friendly or mean? I couldn't tell. Then the kid came by again. He'd been waiting for us, and this time Willie shied and kicked out, and there was a loud noise and—oh my God!—there was a big dent in the rear fender of a black Oldsmobile and the chrome was off. "Oh! Oh!" I thought. "This is it." Nothing happened, though. No one even looked in our direction.

Two blocks farther on, a group of men standing outside a bar turned to look at us, and one of them took off his hat and threw it up in the air and sang out, "Yahoo! The Dixieland Express!" The other men cheered, and across the street there were a couple of family groups on balconies and they yelled over at us and some kids ran into the street and we more or less had to stop if we didn't want to run them down.

"Watch out!" said John. "Don't get near her hind feet."

"He won't bite me, will he, mister?" a girl in a yellow dress said, and as if on cue, Willie let out her mule bray and everyone went scurrying back away from her.

"It's OK," I said. "You can pet her if you want. She likes it." And then they were coming up to Willie and it was nice and friendly and we were feeling good and—to put it mildly —relieved about it all.

We started moving on in ten minutes or so, after answering about a million questions. ("That doesn't take no gasoline, you dummy. That thing makes its *own*." This, from a nine-year-old!) Suddenly, the same kid with the pants came charging up to Willie; but this time instead of hitting her with the pants, he threw them at Willie's feet. Well, Willie bucked and hit out with her feet, and I thought for sure she'd gotten the boy in the head, but fortunately she missed him. John took the boy by the arm and told him to quit or he'd get hurt, for sure, but the boy just grabbed up his shorts and ran around to the front of us again and did the same thing all over. Well, at that, Willie put her ears back and I tightened my grip on her lead rope and got ready because when she did that it always meant trouble. She took a vicious jab at the boy with her teeth, and when he ducked out of the way just in time, it seemed to completely infuriate her. For a minute she was like a statue, and then she reared up on her hind legs and came down on those shorts as if they were someone's head. Stomp! You could hear the sound her shoes made on the pavement a mile away. Stomp! The shorts were in tatters now. Finally, she let out a braying noise like we'd never heard her make before and, ever so carefully, stepped over the rags she'd made of those pants, as if she didn't want to soil her hooves by touching them. Then she was moving down the road at her usual pace, as if nothing at all had happened—or ever would.

You should have seen that boy's face. First his eyes got as big as beach balls. He looked at Willie, and then he looked at us, and then he looked down at what was left of his pants; and then he just ran off down the street as fast as he could run. That was the only "incident" that occurred in Chicago Heights.

We spent the night camped out on airport property and some people who had seen us walk through brought us a picnic supper. We had a ball.

* * *

"If only Kabul Gorge had turned out like Chicago Heights," I think. "But then you can get run down crossing the street, can't you? So don't think things like that. Nothing bad's going to happen to these two Kunst brothers," I say to myself. "Nothing bad's going to happen to them."

CHAPTER

"Go on. Jump across it. You'll see."

Pete takes a run and jumps across the ditch, and before he can walk back to where I am ten Afghans have whooped and hollered and thrown themselves into space; and two of them are down in the sewer ditch laughing and carrying on as if we'd just made up a new national sport for them.

"See what I told you?"

Pete just shakes his head and grins. We'd been watching a pickup game of Buzkashi outside of Kabul. People from the embassy, mostly marines, against some local Afghans. It was pretty boring compared to the real thing so I'd been giving Pete a little lecture on the Afghani mind, as I liked to call it.

But Pete was smiling and having a good time and when he made his run again, clearing the ditch by a good two feet, the Afghans went wild. They mobbed him. They threw him up on their shoulders and carried him around the field chanting something or other and yelling their heads off. Pete's six feet three and about a foot and a half taller than the average Afghan, that's all it is. But these people are like kids. I've tried to tell Pete that, but so far it hasn't sunk in. When the Afghans bring him back to where I am, he's beaming.

That night we have dinner with the mayor of Kabul and

his wife at their home. It's a special gesture to show how badly he feels about John. I can't get over all the attention we're being paid. But just coming into the airport in Kabul brought back everything in a flood and each day we spend here has something in it that reminds me of John and what happened; so I would like to get going with the walk as fast as possible. But everyone insists that we go to dinner here or spend the afternoon there, and then there are all the usual delays—waiting for visas; shots, X rays, checkups; fixing up the wagon; collecting our stuff.

"Where do we sleep?" says Pete. And I explain to him how we lie in the wagon—like two spoons you might say—between the box and the side. Only it's going to be harder with Pete because the space is just six feet long. He looks at me as if I must be kidding.

I show Pete how to grease the axles, where all the gear goes in the box, how to set the brake, and we do a little practice walking with Willie. Willie really looks good after so much rest and good care, and it doesn't take Pete long to get the hang of things, except that right away we're going too fast for him.

"You're just going to have to get used to it all of a sudden," I tell him. "It's our natural pace and I can't change it."

I should let Pete build up his pace gradually. After all, it took John and me a good two months. But the thought of all the time we'll lose if I do that is more than I can stand. Besides, three and a half to four miles an hour *is* how we walk. It's what Willie and I are comfortable with. I probably couldn't even keep Willie back from it. So the first few weeks are a forced march. He was a marine, right? In Vietnam and everything. I tell him to look at it that way.

All he says is "God, what a big mule. Makes Willie the First look like a jack rabbit."

We take about two weeks of supplies from the commissary —thanks to John Turner at the embassy—enough to get us to Islamabad, Pakistan. That's where we'll find out whether we can cross China. He gets us a two-burner Coleman stove, a big improvement over the kerosene stove we started with in Istanbul, home port of our U.S.A.-Turk Machine, home-away-from-home, covered wagon. Also two inflatable mattresses and an ice chest. By March 26, the day we've set for our departure, we're better prepared than John and I ever were. And it's spring. Perfect walking weather.

"These are yours, then?" It's the same police captain that finally took me to the American Dispensary five months ago.

"Yes," I say, and slip the watch back on my wrist.

"It's lucky you had no rings." Colonel Birand's comment flashes through my mind. "They would have cut your fingers off." These are our binoculars, too. I take them and thank him.

"And now, would you like to see the man who killed your brother?" he asks.

Noor Mohammed his name is. They have him in the next room, along with two of the other bandits. The other three escaped into the Khyber Pass. I look at the door and imagine him sitting there.

"No," I say. What would I want to see him for? He would look just like anyone else on the street to me. The captain's brows furrow. This was to have been his great moment. But I don't care.

"What will happen to them?" I ask.

"They will be hung," he says, glowering at me. "Your brother's death will be avenged."

I should feel good, I know, but don't. Just sort of sick.

Going to the dispensary, seeing it all again, is worse than I expected. Dr. Moede's great, though. Jokes around with us in true style. He pumps us full of gamma-globulin shots as well as everything else—for hepatitis; and then, just as we're leaving, he throws Pete a box of pills. "For anything from headaches to diarrhea," he says, and we all laugh.

There is one more ordeal we have to go through—the ceremony in the Kabul Gorge. As we ride toward the spot on the morning of the twenty-sixth, I feel worse and worse. Suddenly, there it is, just as I have remembered it. We stand practically in the same spot where John's blood soaked into the ground, and the governor of Laghman Province makes a speech about how much he regrets John's death. There are a dozen other officials who make long speeches and then there is a press conference. Finally, it is over, and we get Willie out of the truck, hitch her up, and step out at as fast a clip as she will go. We aren't alone this time. There is a Land-Rover full of police leading the way and two motorcycle escorts. The Afghan government isn't about to let anything happen to the Kunst brothers this time. They will be with us to the Afghan border—three days' walk from here.

I mean not to, but as we step out, I look back. I have to sort of say good-bye to John. For an instant, I see him as he was that night in the moonlight, all curled up as if asleep. And then he's standing up and smiling. He's walking along back there next to Willie's tail, the way he always was. Pete's there and John's there, too. I start to belt it out, "Mule Train," in my loudest gorilla voice. One of the guys on the motorcycles turns his head and gives me a funny look, but I don't care. At the second verse, Pete joins in. "It's OK," I say to myself. "It's OK. We came back to get you, old buddy. We wouldn't let you down."

That night we finish our dinner and are drinking a before-bed cup of Irish coffee when we hear a huge commotion outside. "They're putting up some sort of tent," says Pete. I open up the back flap so we can both look. Police are stringing lanterns and setting up tables and chairs under a huge canopy that they have stretched between four trees. Just as they are finished, two Land-Rovers drive up, and behind them a black, shiny Cadillac—an official car.

"Something's happened," I say. "Christ, I hope they're not going to take us back to Kabul."

"They're unloading food," says Pete. "I can smell it."

"Where are the harem girls?" I say. It looks like a scene from some Arab movie. When the feast is all set up, the only policeman who can speak English comes over to us and, smiling like a headwaiter, says, "His Excellency, the governor of Laghman, invites you to a banquet in your honor. Please," and he bows, waiting for us to descend.

Even though we've eaten, we gorge again. What a feast, a real sheikh's banquet. There is chicken and rice soup to begin with; kabobs; pilau with meat, raisins and carrots; fried eggplant; spinach; meatballs with dumplings; lamb's brains; yogurt; and dolma. The governor doesn't speak English, so all we can do is smile and try to show him how much we are enjoying the food. For dessert there is a sweet pudding and piles of fresh fruit! Through the policeman, we indicate to the governor how much we appreciate the honor he has given us. He says he hopes we will remember his province with happiness and will pass through it again someday. Then he leaves. Everything is put back into the Land-Rovers and they drive off. And finally we go to bed. We toss and turn most of the night, though. The feast was great, but it is poor preparation for our first sleep together in the wagon.

The next day we're walking along when I spot a hawk high on a ledge about one hundred yards away. One of the policemen walking with us is particularly proud of his pearl-handled, silver-plated pistol, so, more or less in joke, I bet him he can't hit the bird. I expect him to shrug his shoulders or make some excuse, but he raises his pistol and fires. The hawk flies off, crying. I look over at him as he puts the pistol back into his holster, but he's saying something in Afghan and looking straight ahead of him. The policeman who can speak English translates for us: "It wasn't Allah's will that the hawk be hit by my bullet." Pete laughs, looks over at me and shakes his head. But suddenly all I can think of is John: John lying back there in the Kabul Gorge.

Three days later we're out of Afghanistan and struggling up the steepest mountain road I've ever been on; five miles, straight up, it seems, to the market town of Landi Kotal, at the top of the Khyber Pass. Poor Willie. Her shoes are slipping like crazy. We have to get behind the wagon and push in the steepest parts, which we've never had to do before; and even then she falls to her knees with all the strain.

"What's that?" says Pete, pointing to a plaque in the side of the cliff.

Colonel Khushwant, our new escort, walks over to it. "On this spot," he translates, "by the will of Allah, twelve hundred ninety-six British lost their lives attempting to drive out the Pathan, traditional guardians of the Khyber Pass."

"That's almost a division," Pete says.

"Three regiments," says the colonel. Killed to the last man.

It looks like a picnic spot: a clearing off to the side of the road, cliffs behind it, in a steep gorge, the river below.

"There are many such plaques," the colonel adds. "In 1847 an entire British army was wiped out near here." The colonel, a Pakistani prince from the Chitral, is walking with us through the pass and his presence alone insures our safety. But there are Pathan tribesmen guarding us also. We are the first Americans ever to be allowed the privilege of walking through the pass. We've been warned not to leave the road for any reason. There are tribesmen watching us right now who would shoot us if we did. It's strange to think that somewhere in these mountains, the Pathan are also harboring three of the bandits who killed John.

"The Pathan have a strict code," Colonel Khushwant says, "to give asylum to anyone who demands it, to extend hospitality to all strangers, and to take revenge for any slight. Allah could not find better soldiers." I look over at the tribesman nearest me. He is almost as tall as Pete, very fierce-looking and handsome. How different from the ragged, half-starved Afghan soldiers we had such trouble with. I can't imagine this man allowing himself to be sent off into the desert without food or water or anything to protect him from the cold of the night.

Later that morning we see our first large camel caravan. In the distance it's only a cloud of dust; but then we can see the camels' heads bobbing up and down, their laden bodies rocking back and forth like lifeboats in a heavy sea. There are fifty or so of them, each with a driver. Pete takes picture after picture. I think of the first time John and I saw a camel—just after we crossed into Iran.

"We've got to get a picture of that," John said. A few minutes before, the driver indicated that he'd like to trade his camel for our mule and that had given John an idea. Us standing there holding onto Willie and looking very protective, while the camel driver, with camel in the background, tugged at Willie's bridle. The caption would read: "Kunst brothers won't walk a mile with a camel." It was a great idea. The American press would eat it up, I knew, but there was no one around to take the picture.

"Come on, John. Some other time." I didn't want to stand there in the heat for nothing.

"A truck'll come along in a minute," John said. "I'll flag him down and he can take the picture."

"Or it could be an hour," I said. "Come on." I started to lead Willie down the road.

"Hey, where'ya going?" John had gone over to talk to the camel driver. I heard him yell but I ignored it. What was the point of trying to explain something like this to an ignorant Iranian camel driver?

"Hold up, will you?" I kept walking.

In a couple of minutes John came running up. I could hear his breathing.

"You bastard," he said. "Why are you always in such a fuckin' hurry?" I just kept on walking and didn't say anything. I didn't want to argue with John. That's one thing

I'd discovered. He'd bitch and scream for a while, like he was doing then, but eventually he'd simmer down. When you're together all the time and for as long as we were, the only way to survive is not to fight about every little thing. We got the picture, finally. Later on. Before we left Iran. But it wasn't as good as this one would have been. That camel! He was as big as an elephant, I swear to God.

"Nothing at all," Colonel Khushwant is saying. "The trucks carry everything now." He goes on about the great caravans of the past. The silk trade, the riches of India. Thousands of camels at a time. Alexander the Great and his elephants. Tamberlaine. Genghis Khan sweeping his hordes before him. And as he talks I can see it all: the massacres, the blood-spattered rocks, the cries of the wounded and dying; the acrid smell of camel dung in the dust; the thunder of hooves; the wild shrieking, the whooping of the men as they sweep through to destroy.

Suddenly, behind us, the tooting of an automobile, an ancient Chevrolet. We're walking again now and it barely creeps past us. I count twenty people. Two boys on our side push at the road with bare feet as they pass, as if on a scooter. An immensely fat woman holding a baby goat in her lap occupies the center of the hood. I see the driver craning his neck trying to see around her and the others. Pete and I are laughing. Colonel Khushwant shrugs. This is the mode of travel in Pakistan and India.

About mid-afternoon we reach Landi Kotal, "the smugglers' city." I was expecting something more romantic, but it's just like any other "city" in Turkey, Iran, or Afghanistan: nothing but a big overpopulated village reeking of urine and human and animal offal—with an occasional whiff of jasmine—covered, of course, with a blinding, acrid dust. What a setting, though. We've climbed three thousand feet since entering Pakistan this morning. Way off to the north we can see the snow-covered mountains of the Hindu Kush.

Where we are the terrain is rough and dry, but the foothills to the east are green and all around us are peaks that are capped with snow.

That evening, we get talking with some of the Pathan tribesmen, Colonel Khushwant acting as interpreter. When we tell them we hope to be walking through China, they act very impressed.

"Then you will have to come back here," one of them says. We would be walking through his home country of Gilgit. He points off to the north. Another tribesman shakes his head.

"He says the Karakoram Pass is open only a short time during the year," the colonel translates.

"How high is that?" I ask. Sixteen thousand feet. The mountains around it are twenty-five to twenty-eight thousand feet, the tallest only a thousand feet below Everest.

One man says that to walk around the world that way would be truly a great feat. We would have to leave the wagon behind, of course. They think that if the time is right, perhaps Willie could make it. There are bandits, though. It is very dangerous, in fact. Almost no one goes that way now, only an occasional Chinese caravan, carrying silk and tea.

"So, maybe we won't bother the Chinese," Pete says as we get ready for bed. "Calcutta's downhill from here, right?"

At the very top of the pass is the Fort of the Khyber Rifles. What a sight—the valley spread out below, green and lush in the distance, the road down to it winding and curving through the desolate hills. "What a place to live," I think. "All of India and Pakistan at your feet." We keep up a good pace for we want to be in Peshawar, twenty-one miles away, tonight.

About two in the afternoon we are cut off from the others by a flash flood. All at once there is a rumble and then, without warning, a muddy torrent boils across just in front of us and spins off into the air. There are six tribesmen with us and they immediately sit down and take out cigarettes. An hour, two hours, they say. I do not want to wait that long, though, so after maybe thirty minutes, when the waters seem to have calmed down somewhat, I lead Willie Make It out into the stream and slowly move across. Water boils at my knees. Halfway across Pete yells out, "The wagon's slipping."

"Hold it," I say. "Just stay where you are." The wagon comes to rest. We are at a thirty-degree angle to the stream now. To my left is the cliff's edge, nearer to Pete than to me, the water shooting over it like out of a giant hose. "OK. Slowly. Hold the wagon all you can." I pull at Willie's lead rope and we move ahead. Slipping a little more to the left with each step, we still move forward, into shallower water, though; and, finally, onto dry roadbed.

"Whew!" says Pete. He sits down and dumps his shoes.

"We could have used ropes if we'd really been worried about it," I say.

"The next time, let's do that, shall we?" We sit for a minute, resting. Colonel Khushwant and the others have been watching us without saying a word. Now they get up and make ready to move on again. As suddenly as it came up, the flood subsides. Now the Pathan are bounding across the stream like goats. We are walking again, at a good pace. But our shoes are wet and soon we develop blisters. Son of a bitch, I say to myself, as I feel them coming on. Son of a bitch.

At the bottom of the pass, but still seven miles from Peshawar, is the mud fort of Jamrud and the arch that marks the end of the tribal territory. There is a huge crowd there to greet us: Mr. Velletri, the American consul; the political agent for the Pathan; and about one hundred tribesmen. There is a brief ceremony in our honor, after which Colonel Khushwant leaves us. For a man of sixty he is in excellent shape, but he admits to being tired. I can see the pain in Pete's face as we pick up the walk again, but we don't say anything, either of us, and in a couple of hours we are in Peshawar.

Luckily, we are staying at Mr. Velletri's, so we are able to have a hot shower before we go to bed. No matter what the day is like, if we can end with that it's not so bad. That's my feeling, anyway. John and I went twenty days without a shower in eastern Turkey, I tell Pete. And the temperature was well over one hundred most of that time. Pete just groans. His feet are so sore he has barely been able to make it these last few miles. I don't know how he's going to walk at all tomorrow. Still we're here, aren't we? Pakistan. Another country. Only a little over a week to Islamabad and to what the Chinese will say.

CHAPTER

It's called the Grand Trunk Road, don't ask me why, and it's the worst road we've come to yet. It's the only one straight across Pakistan, though, so we have to take it. The surfaced part is bad enough, full of potholes and hardly any of it really smooth, but off of that it's unbelievable: like walking through a construction site where it hasn't rained in about a year. You choke in the dust and stumble over rocks, and in general it's so terrible that, no matter what, we try to stay on the paved part. Luckily, there aren't that many trucks or fast cars. It's only really bad when we come to a bullock cart or something. Then there're two of us abreast, so if something comes in the opposite direction, watch out!

We've spent two days in Peshawar because of Pete's feet, but on April 3 we move out. He doesn't know what it's like, I keep telling him. He could wait forever, and his feet and legs would still hurt him. He's just got to take it, suffer it out the way John and I did. Eventually they'll toughen up; they've got to. The fourth day out, though, Pete tells me he can't go on if I don't let him rest. "Some marine you turned out to be," I say, and keep on packing. But Pete just lies there in the wagon, reading, as if he hasn't heard me. "What the hell did you come along for anyway, if you can't push it out," I say, good and nasty. No response.

I go over to where he's lying and accidentally on purpose knock the book out of his hand. "Whoops," I say, and kick it under the wagon. "Now how did that happen?" John would have been all over me by this time, but Pete just lies back down flat on his back and lets out a long sigh.

"Got to take the day off, buddy. Feet sending me messages. So the world walk takes one day longer." Pete leans up on one elbow. "Give me back the goddamn book, will you? Like a nice brother? Or I quit!"

I get the book for him. What's the use?

We're lying there reading, or rather Pete is—I'm too nervous, and irritated—when we see a great crowd of people coming upon us from the direction of Peshawar. One of the reasons I didn't push too hard earlier was that for once there weren't any people around to speak of. Whenever that happens, my motto is enjoy it all you can because it's not going to last forever. Well, here they are. Hundreds of them. Thousands. At first it looks as if they might leave us alone, but then they begin to turn off, a few at a time, and pretty soon there's a whole crowd of them around the wagon; they're circling us, pressing up closer and closer, and I know in a moment there's going to be trouble.

"Hoo, ha, ha, ha. Hoo, ha, ha, ha," yells Pete and they really let loose, start hooting and hollering just like a pack of savages. Pete loves it. He's doing the Indian business louder than any of them, and all the time they get closer and more and more of them join the circle.

Then it happens. One of the men holds up a bottle of something—perfume it looks like. He wants Pete to buy it, of course. Well, Pete just takes it and puts it behind him in the wagon and waves to the man. He's thanking him, the dummy. He thinks he's being given a present. So the man grabs Pete by the arm. He shouts at him, and you can tell by his face that he's mad as hell. He wants to be paid, of course. Pete just laughs and grabs the man back, but you don't playful wrestle with these people; Pete ought to know that by now. The man lets out a yell as if he's been murdered and jumps back, and I'm terrified he's going to pull a knife. But he doesn't. He just shouts something over and over again very loud, all the time looking more and more angry. And then the mood of the crowd shifts. You can feel the hatred like a blast of hot wind. Suddenly, they're quiet. They've stopped moving, too. What is the stupid dork saying, anyway? It sounds as if he wants to kill us.

All this time Pete has been trying to calm the guy down, but everything he does seems to anger the crowd. Then he reaches back and grabs the bottle and tries to hand it back to the man, and there're other angry faces and other dorks are yelling, and there're fists waving around, and in a minute they're going to push over the wagon and take all our stuff and probably beat us to death. But by this time I've reached into the provisions box and found what I want. It's none too soon, either, as I can feel the wagon start to rock back and forth. I scramble over to where Pete is at the back of the wagon and, crouching there where everyone can see me, I point the camera at the peddler, click off a couple of pictures, and then gesture from the man to the camera and back again and yell "Police! Photo to police!" very loud and clear a couple of times. Well, the guy looks at the camera for a second and then up at me and then he turns and pushes his way through the crowd and in a minute he's running down the road toward Peshawar as if there are mad dogs on his heels. Then the whole mob disappears. Suddenly they're gone, and here we are alone again. Pete just looks at me in disbelief and I laugh.

"Whew," says Pete, and sits down on the box. He's pale as a glass of milk, his mouth still sort of trembling. I nod my head and grin and put the camera back in the box. John and I learned that trick in Iran; but I don't want to let on to Pete how many times it doesn't work, so I just keep my mouth shut. I could have used it sooner, or I might have thought of something else to stop things from getting that bad. The truth is, though, I wanted Pete to see what it was like when these people get out of control like that. I'd shown him all right. I'm just glad it wasn't something worse.

About two o'clock we hitch up Willie and move out. It's twenty-five miles to Islamabad, and if we want to make it by tomorrow night, we'd better do some of the walking today. Pete suggests it himself. It's right in the worst heat of the day and I would just as soon wait and do it all the next day, but I think that resting by the side of the road has lost some of its charm for brother Pete. Anyway, I don't object, and later that afternoon some nuns who are driving by in a panel truck stop and talk to us and they end up by inviting us to stay in their convent for the night, and for the first time since Peshawar we have hot showers and a home-cooked meal and good beds to sleep in and it sure feels great. They're European nuns and very interested in our walk, and when Pete tells them what happened they say we are very lucky, and I get even more credit for the camera trick.

On the way into Islamabad the next day, the crowds really get to Pete. "What are you doing?" he keeps shouting. "Get your hands off of that!" he'll say, as someone reaches in and grabs a pot or something. "Hey!" he yells at a boy who's inching off with one of our plastic water jugs. He runs right after him and the boy drops it, but I warn Pete about leaving the wagon like that. If you run off by yourself into the middle of a mob, you never know what might happen. "Take it easy," I tell him. "Use that switch if you have to, but don't chase after them." I know how he feels, though. Sometimes these people can make you so goddamn mad you don't know what you're doing.

It was a little village in Turkey, up in the mountains,

and it was cold and we hadn't been able to get any good feed for Willie, so John and I stopped and tried to see if we could buy some. Usually, we'd walk through these villages as fast as we could. We'd be walking along and a boy would throw a rock at us. If we kept going, maybe there'd be a few more boys throwing rocks and maybe we'd even get hit by some of them; but if we stopped, right away there'd be a whole crowd of men and boys and pretty soon just about anything could happen. It was as if they were all just sitting around waiting for us to come along. The first time it happened, we couldn't believe it; we'd hardly crossed the border and right away it was like being in enemy territory. In Bulgaria the people hadn't bothered us at all. They seemed too scared to do much of anything. It was the police that made us feel nervous there; but in Turkey it was just the opposite. Even the police were scared of the people, we found out.

Anyway, I happened to see some hay piled in the corner of a shed, so we stopped while I tried to explain to the people that we wanted to buy some. The thing we could never get over was how stupid they were. We'd get the words all right: "Hay," "sell." How many did you need, anyway? Then, usually whatever price they'd say, we'd offer half, sometimes less. Believe me, there wasn't one of them that wasn't smart as hell when it came to gypping you for all you were worth. This time I thought it would be easy because there the hay was. All we had to do was find the man who owned it. Well, we found him all right, but he wouldn't sell it to us. He just looked at us as if he wanted to kill us, spat at our feet, and walked off.

They weren't all like that, but there were always some in these countries who just hated foreigners, Americans in particular, it seemed. So we moved on. We were going to walk right through the town and out the other end and just find what we could for Willie along the road or give him some bread we had, but the villagers wouldn't let us.

They got in front of us so we couldn't move, and then they pressed up around us and started to steal our stuff. Once one of them got something, it made it twice as tough because then they all thought they could get away with it. Then some dork picked up Drifter and started to run off with him. Drifter II, that is. No Turk could have gotten near Drifter I. That drove John nuts, when they teased Drifter, so he went tearing off after the guy, leaving me to handle the rest of them by myself. So I just led Willie right through them as fast as I could, and prayed that John was going to be able to take care of himself. Luckily, no one got caught under the wheels and Willie didn't bite anybody and after a while I was on the outskirts of the village and the people were beginning to thin out a little and then John came running up with Drifter in his arms and I thought that everything was over.

But it wasn't. John was still furious, so when some kid threw a rock and it hit him in the neck, he turned and ran after him and I had to stop. In a minute they were pelting us with rocks and we were both hit, and then we went berserk and charged them. Well, some of them turned and ran when we did that, but there were about a dozen men who just stood there. We were right on top of them, and then instead of their hitting us or running away, they stayed exactly where they were, their arms at their sides. It was as if they'd suddenly gone blind. They just stood there, like a human wall. We could have hit them. We could have killed them probably, but instead we just came to a halt ourselves; and then, when all they did was keep those shit-eating smiles on their faces, we turned and walked back to the wagon and moved off again. They started throwing rocks again, but this time we kept moving. It was the damnedest thing we'd ever seen. We couldn't figure out what was in their minds. Can you imagine just standing there stock still when someone comes rushing up with blood in his eyes? It was some sort of stupid game, we finally decided, though what it meant or what the rules were we never did figure out. Anyway, after that, before we'd walk through one of these villages, we'd brace ourselves with a snort of whiskey, or whatever else we had, and we'd never stop unless we absolutely had to.

As we get into Islamabad, things start getting better. Islamabad is the capital and a completely new city. All the government offices and embassies are there. The people all live in Rawalpindi, the old city just beyond it. It's the cleanest, best, most modern city we've seen on the whole walk. We even find a private campground: a grassy place with a fountain and some benches and *no people*. We've eaten and are sitting around resting and looking out at the sunset over the city, enjoying the relative coolness and most of all the fact that there aren't any people around, when all of a sudden we hear this American voice: "Hey, you guys! You the Kunst brothers?"

I turn my head. "Yeah," I yell out.

Blue-eyed, blond-haired Jack Armstrong, the all-American boy, yells back from his Volkswagen window, "They're expecting you at the embassy. Where have you been?"

CHAPTER

DAN SCANLAN, OF ALL PEOPLE, IS BUYING US DRINKS. I HAVEN'T seen him since the eighth grade and here he is working for some oil company out in the desert, and here I am, hotfooting it around the world, and we bump into each other, literally, in the bar of the American Club in Islamabad, Pakistan. Small world, as we keep saying over and over. Dan and I are filling each other in on the last twenty years and Pete is looking around the room, his smile getting broader with each drink. In a minute or two we're going to have a couple of big, juicy steaks on old Dan.

The embassy's got us an apartment, Willie's taken care of at the Islamabad Club Stables, and now we've met Dan, who's already insisted that we charge up all our meals and drinks to him. So it looks like a pretty good couple of weeks coming up. We'd planned on only one week, but what the hell! We need that time to figure out our best approach to the Chinese.

Pete's dancing with some chick and I'm looking around for my little Pakistani secretary, and all the time Dan keeps yakking away about what a coincidence it is that we've met again like this.

"It's fate," I tell him, and he smiles and nods and keeps on about it, but I'm serious. More and more I have the sense that what happens to us happens for some reason, that it isn't just chance. I'm on the walk because I'm meant to be on the walk, for instance. Sometimes I even get the feeling that I've done it all before. Places that I've never seen look familiar. We're walking along down a street, say, and I *know* that I'm going to see a certain restaurant sign around the corner on the right, halfway down the block. And there it is. That's happened so many times I don't even get surprised anymore. It's like when my friend Rich told me, "Why don't you walk around the world? That's never been done before."

"It hasn't?" I said. I knew right then and there I was going to do it. Hadn't I said that the day I was thirty I was going to quit the survey crew and do something else? And hadn't the walk come up just five months before that? Fate! Sure it is. No one can tell me any different. Every time I get to the shooting, though, my ideas seem to go all to pieces. John killed and me not? Why? He'd moved, the captain told me. So what? Maybe he couldn't help it? But maybe it wasn't meant to be fair, either. I don't know. Sometimes I think life is like a great, big bowling alley. There are millions of pins, millions of them, and God just throws the balls and even he doesn't know who they'll hit. Things happen because they have to happen, that's all. It's not chance, though. There's some reason they happen the way they do. It's just that we can't figure it out all the time.

The next day is April 12. It's a big day because finally, after twenty-five years of haggling back and forth, the Pakistani constitution is going to be signed. We've been here four days, so we get included in the celebrations. Mostly they're parties, of course, but we do get our picture in the paper with the minister of tourism, a Pakistani prince. We wanted to get Willie in on the act, but the Pakistanis didn't like that idea. It wouldn't be proper. Just like with the Shah of Iran: too big a shot to be seen with a lowly mule. We'd gotten a couple of American generals to pose with

Willie for the papers, instead. Fixed it so it looked as if Willie were laughing at them. They got a big kick out of it. But we hadn't even been able to see the Shah, much less see him with Willie.

We're lolling around the pool at the Canadian embassy when Ben and Carol Bennett come by and ask us if we want to go to the bazaar with them. They're with the American embassy and good friends of ours. Well, I'd just as soon stay where we are, but Pete says, "Sure," and jumps up, so I go along for the ride.

This bazaar's like all the others, only it's bigger and smellier and hotter, it seems. It's like one big flea market, actually, only with real fleas and plenty of spooky-looking characters. As soon as we get out of the car, this beggar comes up to us. He's got only one leg and his right arm is twisted all up like a turkey wing, and Pete, like a dope, gives him a coin before the rest of us can stop him, and then we're practically mobbed. Our beggar's knocked down, though it doesn't seem to bother him any, and we almost have to give it up and get back into our car, but then Ben, who knows this place like the back of his hand, gets us away from them and suddenly we're looking at a lot of brass pots and Carol is haggling with a man about one of them. We wander from one place to another, getting deeper and deeper into the bazaar, until I've lost my sense of direction entirely and am beginning to get nervous.

And then it happens. Right in front of Carol. A man pulls a knife and yells something and lunges at the man next to him. The other man crumples to the ground. I hear him cry out over and over again—it's like the sharp wailing of a dog—and then he stops and we see a pool of blood spreading toward us from under his legs. Pete touches the man on the shoulder. He falls to the right and over onto his back, and then we can see the blood pumping out of his neck and his white, pupilless eyes that are rolled up into his head. He lies there, twitching, his body arching up into the air in sudden convulsions. Carol just backs away from it all toward Ben, silently and quickly. And there is the man who did it. He hasn't run away into the crowd. He's standing there not ten feet from me, jabbing his knife out in front of him and crying out something in a loud, piercing voice. Pete's over on the other side of him, backing away, I'm glad to see. I start to back off, too, when suddenly the man runs forward and cuts into the face of an old woman who is pressed up against a counter on my left. At the same instant I hear her cry, I see him wheel and cut blindly behind him, slicing the sleeve on the arm of a small boy who is hugging a basket of fruit to his chest. The blood leaps to stain the white cloth, the boy's face goes blank; there are a dozen cries, shouts; a woman's voice wails above it all. Then, suddenly, there is quiet. Absolute quiet, as if a switch has turned it all off. The man is crouched in front of us, poised, ringed in by the narrow street and the people, who do not move a muscle. Then he lunges again, this time at a man who slashes back at him with his own knife. The spell is broken. There are thirty people on him now, half of them with knives, it seems. We retreat through the thinning crowd, down a tiny alley that reeks of urine, and then out into another area where the people are jammed in on our left, where there is another fight, maybe, or where one of the victims of the maniac killer has stumbled. We run to the right, where there is an opening, and in ten minutes or so we are back at our car. Thank God for Ben, who does seem to know this place after all.

We get in the car and drive off. Round a corner we almost run down an old man who is praying in the street, part of the overflow crowd from the gigantic mosque that is in front of us. It is five o'clock. The whole city is at prayer. We sit in the stalled car, sweltering in the heat, waiting for it to be over. Finally, we can push through the crowd and we're on our way back to Islamabad. The Bennetts drop us off at our apartment. We are too shaken, all of us, to want to go anywhere.

A week later we're talking with a Mr. Sing at the Chinese embassy. We've tried a dozen different ways of approaching the Chinese and have finally been advised to go there ourselves and ask. "Ah," says Mr. Sing. "David and Peter, the two American brothers who want to walk across my country with two Chinese brothers! My country very big. Might take the rest of your lives to walk across my country." He motions us to sit down. He is smiling broadly. His cheeks are red as apples and the top of his head looks polished. He is very amused but also very polite, so we don't mind so much that we're being refused. We hardly know, in fact, that we are being refused. There are bandits, he says. The terrain is among the most difficult and dangerous in the world, particularly in the western part. "We would not be able to watch and care for you as we would wish," he says. "No, it is much better that you not walk through my country at this time. Someday, when there are no more bandits. Someday . . ." He drifts off. We sip tea. We talk of our travels thus far. It seems out of place to remind him that the Karakoram Pass is still a trade route, that if others have crossed the mountains, probably we can, too. We've about given up anyway. Also, as Pete said, it's downhill all the way to Calcutta.

It would be the teeming masses all right, but we've already made it part of the way. How much worse can it get? As John used to say, no matter where we do it, it's just one step at a time. Anyway, we've come up with another idea in the last few days that makes us feel almost enthusiastic about the India route. If we aren't going to make a perfect circle around the world, then why not drop down to the Southern Hemisphere and walk across Australia? Crossing Australia will be a little longer but we'll get onto another continent and there's one other very big advantage—we'll know the language. It'll be like walking across the United States.

Once we get the word from the Chinese, I want to start right away, but this time it's the Pakistani Tourist Bureau that holds us up. They want to take us up to Chitral, in the Swat Valley area of northern Pakistan: "the Switzerland of Pakistan," it's called. Once again Pete is all hot to go. "Anything to get out of walking," I kid him by saying. The trip turns out to be one of the high points of the walk for me, though: flying, flying in a small plane in those huge mountains! It's as if the wings are on me. Pete and the two other guys from the embassy who are with us are airsick, but not me. Every time the plane drops, I drop with it, my whole body. We swoop into a valley, rise like a bird up over the opposite ridge, and glide down into another valley more lush and beautiful than the one before. Crystal streams dazzle our eyes. The glaciers of the Hindu Kush rise impassive and awesome in the distance. We're at the top of the world. When we get out, the others have problems breathing in the rarefied atmosphere, but I just feel lightheaded, light, too, weightless that is. Then one day we take a jeep up into a hot springs valley just under Mount Tirich Mir, which is only one thousand feet lower than Mount Everest, and I don't ever want to come down again. It's like *Lost Horizons*. There's no monastery or anything, but we're in a tiny valley that is a true bowl set in among the highest peaks in the world and it's hot, tropical almost, and there's water everywhere and as I look around all I can see is the bamboo and other heavy vegetation of the valley, while above are barren rocks that look as if they've just been made, and above that snow and a jagged circle of peaks, and nothing else except the blue, blue, blue of the sky. "Someday I'll come back here," I say to myself, as we leave. "Someday when the walk's over, I'll come back here and live."

Finally, we're leaving. It's late on the morning of May 16 and we're standing there steaming in the light rain, waiting for the media to finish with us. Pete has on galoshes and rain gear, but I'm dressed in my usual short-sleeved shirt and long pants. My hat always gives me pretty good protection, but what's the difference? We'll be soaked with sweat soon enough anyway. If John were here, he'd probably have his shirt off.

The road is even worse than before. Every time a car goes by we're splattered with mud. If we get off onto the shoulder, Pete has to get back behind the wagon and push. Pete can't keep up with us with all that gear on—but he had to try, didn't he?—so he takes it all off and pretty soon we're both soaked, feet and all, and covered with mud, and it's so hot that when we walk through the villages we try not to slow down even, because the press of people around us cuts off all the air, it seems, and it's like being in a steam bath. What really gets to Pete, though, is the trucks: big semis that come barreling along every half hour or so and don't slow down for anything. Pete's going crazy with them—shouting and shaking his fist and pulling us off the road every time one goes by. But I remember Spain, as well as Iran and other places, and in comparison this is nothing.

It was about five miles from Madrid. A truck blared its horn and we got down into the ditch. He *had* to get over. Another truck was passing him. An oncoming car weaved back and forth and finally went down into the ditch on the other side. Whew! Made it! Big ruts in the soft shoulder right where John and I'd been walking.

We got Willie back up on the road and pushed on again. At first we'd tried to fake them out, the way we had in the United States when it was a narrow two-lane. Let the traffic back up behind us. What the hell. But here we couldn't do that. These guys were crazy. They'd pass us without even slowing down, and if there was a car coming, they played dodge car. The first time it happened, the car passing us nicked the pack and Willie spooked, and we ended up down in the ditch with me yelling at them and swearing. The next time I moved over a little but not enough, and we ended in the ditch again. The third time John screamed out and we stumbled down into the ditch before the truck got to us. A close call, as he was over onto the shoulder himself when he went by.

"Watch it!" shouted John. Coming right at us were two trucks. There was a car behind us that we'd managed to hold up for a minute—probably an American tourist. No way could they all pass. I looked up from the ditch expecting to see a head-on, and would you believe it, another truck was passing the car that was behind us. In about a second that would be the end of all four of them. And us, too, probably. For the first time on the walk I was scared. Scared with me is when I get this cold feeling and then it's as if I'm watching a replay of a movie I didn't particularly want to see anyway. Everything sort of slows down and suddenly I'm looking at myself as well. There are two me's. One's up above me someplace. About eight feet above my head, as a matter of fact, and it's looking down at the other me and at everything and kind of smiling. It's weird, I know. It's just the opposite of what most people would call "scared." But if I had hairs on the back of my neck, they'd be standing up. I know I'm scared because afterward I'm shaking all over.

Well, it turned out like in one of those old-fashioned movies. The car swerved off the road halfway down into the ditch just in front of us. The oncoming truck did the same on his side. And the two passing trucks smashed sideview mirrors off each other. There wasn't an arm's width between any of them. Unbelievable. I hadn't seen anything like it since the stunt drivers at the state fair. Spanish drivers! You've got to hand it to them, though. They've got balls.

We've been told to camp for the night in the villages. It isn't safe to walk after dark or to stay out in the open. What they don't tell us, though, is that we don't get any sleep if we do that. The first night out we literally stand guard the whole time. And if we aren't actually pushing people away from the wagon, or grabbing a bony wrist or a hand with a can in it, it's so stifling hot in under the canvas that it's impossible to sleep. What we do the next night and the ones after that is what John and I did through all the other countries. We walk until we're too tired to go any farther and then just pull the wagon off the road anywhere and crawl into it. If the bandits get us, they get us, that's all. We try to get moving by 5 A.M. or so, too, to get the jump on the dorks. The villages are pretty close together and it doesn't take long for the roads to get crowded. In some places there's no space at all in between them. That's where it's rough. "People, people everywhere and not a place to think." Pete's song. He's beginning to sing it all the time now. We both are. You can't throw an empty tin can out on the ground without a dozen dorks fighting each other to get it. Luckily, we have everything we need in the wagon and don't have to stop, except every couple of days for feed and water.

Two days this side of Lahore, we're resting on a little bridge near a farmhouse because Pete's feet are sending him messages again, and over hobbles an old man. He looks just like all the other beggars, and it's the same scratchy, poor country with not a machine in sight, so we brace ourselves. It's instinctive now.

"Hello, sirs," he says. You could have knocked me down with a feather. "You like plenty these?" What he's got in his hands are some beautiful-looking, red tomatoes.

"How much?" says Pete.

"For you. Present for you, sirs," the old man says and smiles so broadly you can see where all his teeth used to be.

"Thank you very much," I say and take them from him. They're delicious: juicy and ripe all the way through. I throw one to Pete.

The old man is patting Willie now. He has to lean on his stick and stretch way up to touch Willie's head and I think for a second that Willie's going to swing at him, but it's OK.

"What would I give for her, sirs," he says, and looks over at us if he's just seen the sight beyond compare. "More beautiful than all my daughters. Stronger than all my sons."

There's a donkey over by his house. A camel stands splay-legged in the dirt-packed yard. The old man raises his hand and beckons in the direction of the house, and immediately three boys and four middle-aged men come running down to us. I see now that they have been sitting on the family char-poy, the rope-mattressed bed that sits in the shade in front of all the houses around here. The women, of course, stay where they are. If Willie didn't create such a fuss every place we go through in Pakistan, I would have been worried at this point, but I know now it's the same impulse that leads us to gather around a new Rolls Royce on the block. They bring Willie some chaff. We eat some more tomatoes.

Two of the men speak English a little, so we learn something about their lives. One of them has been to college. All would like to go to America. It is a treat to see a Pakistani smiling and enjoying himself. They seem very happy. Maybe it's the tomatoes. Maybe it's Willie. Who knows? Anyway, it's the first conversation we've had with anyone outside the embassies since we left the Pathan, and that was through an interpreter. They want us to spend the night but it's still early in the afternoon, so after about half an hour we start off again. One of the little boys runs along with us when we leave. "Williemakeit!" he shouts, and squeals with laughter. "Williemakeit!" Other people on the road smile and look over at us. We're sorry when he turns and walks back in the direction of his home again.

We've just started walking again the next day after our noon rest when a car stops in front of us on the other side of the road and a man gets out and walks over, smiling. He's got a big bucket with him which he sets down in front of us.

It's an ice bucket full of Cokes! Wowee! Before we even find out who he is or what he wants, we down three ice-cold Cokes apiece, and then in between laughs and headshakings, he tells us that he is from the Pakistani Tourist Bureau in Lahore and is there to prepare our welcome into the city on the next day.

Things are looking up. He wants to meet us with representatives from the media and some government officials at a certain bridge at nine in the morning, so we say fine and drink the rest of the Cokes—he's brought a dozen and we can't believe how good they taste in this sweltering heat—and when he leaves he smiles and promises us more Cokes in the morning. Boy, do we feel good now. We're going to get the VIP treatment for sure, this time, I tell Pete. We haven't had a really cold Coke since Islamabad. They have them, all right. But usually they serve them warm. Hot, actually. A hot Coke. Imagine! It's amazing how you miss things like that. Like with the guys at the marine base in Turkey. One of them got a package from a buddy who'd been flown back to the States. We were all in the mess hall when he came in and opened it up. What it was was some McDonald's hamburger wrappers. We passed them around like holy objects, so we could each take a good, long smell.

The Pakistani Tourist Bureau officials in Lahore are as good to us as any we've come across so far. They all speak English and they take it upon themselves to show us around what is probably the most interesting city in this part of the world. For three days we're just like regular tourists, but the fact that our guides speak English is what really makes the difference. After we're shown the famous, green bronze cannon ZamZama, I'm dying to read *Kim* by Rudyard Kipling. I can't get a copy, though. We spend a morning at the Shalimar Gardens, which were built by the Moguls for royal family outings. They look a little like pictures I've seen of Washington's Mount Vernon. The Lahore Boy Scouts hold a big banquet for us there another day, which turns out to be pretty funny when Pete falls into one of the royal pools chasing the royal fish. What I get the biggest kick out of, though, is the Lahore Fort. We spend hours going through it and for once I can ask all the questions I want to. Then our guide gives me a book, mostly about Akbar the Great, which I hardly take my nose out of for the next two days, and we come back again. We walk the huge elephant steps that lead right up to the palace, and I imagine what it must have been like to be king of all this, to be living way back in the sixteenth century and be able to get people to do things like this for you for no other reason except that you ordered them to. Why should you walk if you don't have to, right? You're the emperor, after all, and what you say goes. The night before we leave they put on a light show for us at the Lahore Fort that makes our Fourth of July look sick. Boy, how I wish every place we went to was like Lahore.

The American consulate stocks us up with supplies and promises to have twenty-four gallons of good water ready for us on the day we leave. Willie's been shod and is fat and sassy, and the wheels of the wagon have been greased and tightened. So, finally, we have to leave. We sure don't want to. What's coming up next is India—1,300 miles of it. And it's the hottest time of the year.

CHAPTER

"So where is everybody?" says Pete.

"Wait till we stop," I say. But I'm surprised, too. The border hadn't been too bad. For some reason we'd expected there to be masses of people crowding the roads as soon as we hit India. In fact, not only is there practically no one around, but the countryside is getting prettier—less desert-like. There are plants and bushes alongside the road, even an occasional tree. We stop for lunch in the shade of a clump of trees and while we're eating, what should come strutting over to us but a peacock.

"Look at that tail," says Pete. It's the first one I've ever come across outside a zoo, and seeing it here surprises me almost more than there being so few people. Maybe we've got it all wrong, I say to myself. Maybe we've been through the worst of it. We move off after only half an hour because we want to make it to Amritsar before the tourist office closes. Just five days in a hotel have softened us up all over again. Spoiled Americans. That's us. The only problem is the mosquitoes. They're the biggest I've ever seen and they come in clouds. There'll be none at all and we'll be feeding the peacocks, or just lying down on the grass enjoying the shade and the quietness, when suddenly they'll hit us. "Zzzzzzzzz." We dive for the wagon, but they get us there, too. So we push off even faster than we'd planned, and on the road they're not so bad.

The tourist bureau in Amritsar is open but the guy in charge can do nothing for us. They don't listen to him in New Delhi, he says. He's asked for all sorts of things: maps, a car, so he can take people around. They tell him he's lucky to be getting paid. He says we can park our wagon in the tourist office yard for the night. That's the most he can do. We spend a terrible night swatting mosquitoes —maybe that's why there're no people—and the next morning we can't even get off early the way we want to because he insists on showing us the Golden Temple, the most famous of all the temples belonging to the Sikhs.

Well, having him along to explain things sure makes a difference, I have to admit. I'd almost gotten used to the idea of walking through countries and never looking at anything because what was the point if we couldn't talk to anyone about it or read anything. There were so many places like that in Spain, for instance. Even though John knew a little Spanish, it didn't seem to be enough, and if we wanted an explanation of something we always had to shell out. And we sure didn't have money to spend on that. It's a pretty interesting place, though, the Golden Temple. We have to take off our shoes and walk through some water to get into it, and while we're wandering around the tourist guy tells us all about the Sikhs and their way of life.

"A Singh is always a Sikh, but a Sikh is not always a Singh," he says, and laughs. "Singh" means "lion," we find out. And the Sikhs are famous fighters. They also seem to have some good ideas—like believing that everybody is created equal and that Hindus and Moslems shouldn't try to kill each other all the time. No one pays much attention to them, naturally, but that's par for the course, I guess. There are some Sikhs standing around, and both Pete and I are impressed about how strong and happy-looking they are. Unlike most of the people we've seen in this part of the world, they actually look as if they had a good, square meal within the last day or so. They're all dressed in clean, white, baggy pants and shirts and they all wear turbans and have beards. They aren't supposed to cut their hair. That's one of their beliefs; don't ask me why. It's also the reason for the turbans. It's a nice temple, too. There's a large artificial pond and a little bridge that goes across it and there's music playing and a guy reading aloud from the holy book and in the back a kitchen where they dole out food to anyone who needs it. The whole Punjab is Sikh territory. They have more education than most Indians. We won't be bothered by them along the road, the man assures us. How far does the Punjab extend? Until New Delhi, three hundred miles from here.

It's very hot and the mosquitoes bother us a lot until we rig up something bearable with some netting we finally get; but each day that we walk, the countryside gets more and more beautiful, and it's true, the people don't pester us at all. It's flat country, farmland, and very fertile. It reminds me a little of Indiana, though of course they don't have the huge cornfields or anything like that. What makes me think of Indiana, though, is that Pete's complaining about how hot it is, and I remember what it was like for John and me and realize how much I've gotten used to things since then.

We'd been looking forward to Schererville because that's where Route 30 turned four lanes, right through Indiana and Ohio. Flat and straight. Good walking.

Wrong! Terrible walking! For one thing, as soon as we crossed the state line, it climbed to one hundred degrees, ninety percent humidity. By eight in the morning we were soaked to the skin and that's how we stayed all day. We wouldn't even dry out at night; just get up and put on the same stinking shirts. Worse than that, though, we didn't have any sense we were getting anywhere: no people, no towns, no gas stations, even. Nothing but a black, sizzling treadmill of stinking tar through the corn. Corn! We got so we'd look forward to the potato fields because at least there we could see out over them.

Schererville to Valparaiso: eighteen miles. We camped at the exit where there was some grass. I had to walk a mile and a half to find a gas station where I could get Cokes and some water for Willie.

The next day we walked only fifteen miles. When we saw a motel right off the highway on Route 39 we just made for it; didn't even bother to call the papers or try to get it for nothing. The guy was interested in our trip, too, but we hardly talked to him. All we could think of was to get in to where it was cool. Hit the showers. Lie down on a bed and watch TV.

The trouble was we were spoiled. Here we were in our third state. We'd been walking for a month and had covered more than four hundred miles and this was the first time we'd paid for our lodging. We would have done the same thing if it had been our last buck, I think. You see, when you're walking on a two-lane road through every little town, people stop and talk to you. You get a Coke. You wave at the cars. Through a city like Madison, Wisconsin, for instance, it was fantastic. Some of the people had never even seen a picture of a mule. Others couldn't believe we were really out to walk around the world. People were always asking us to their homes for a meal, or a shower, or to spend the night. There were all kinds of things going on all the time and we were at the center of it. After a while, it got to be old hat, but the first time some girls stopped and asked for our autographs, John and I nearly flipped. On the four-lane highway, though, we might as well not have existed. Nobody knew about us and nobody cared. We even stopped waving at the cars.

In spite of the heat we're making good time, about twenty-five miles a day. I want to do more, Pete less, naturally; so we compromise. His feet are still killing him, poor guy. And he's tried everything: thin socks like I wear, thick socks. Boots, Redwing oxfords—policeman shoes—what John and I always wore. They're the best. But nothing's really good for him. How the hell did he make it through the marines? I ask him. "The longest march we ever took was ten miles," he says. "And that was in boot camp." What is it, the whole world's going soft? What the mailmen told John and me was right for us—three weeks for your blisters to turn into calluses—but it doesn't seem to be holding true for Pete. Oh, they're not as bad as they were. But they're still giving him a lot of pain. And his legs hurt him. He gets cramps in his calfs during the night. And then, of course, he wakes me up. And I tell him to shut up. And that gets him mad. And we yell at each other. And some nights I get so mad I jump out of the wagon with my bag and crawl in underneath or just lie out under the stars. John was a lot better to sleep with than Pete, that's for sure. Pete keeps talking in his sleep for one thing, mostly about his wife and kids. He misses them, goddamnit. So, I miss mine, too, I tell him. He's bonier than John was. That's another thing. Of course, John and I had more time to get used to each other before we had to crawl into that sardine can and bunk together; but still, John was easier to sleep with. Sleeping was one of the things he did best, as a matter of fact. I used to curse him for that talent plenty of times, but I have to admit, it had its advantages.

It's six days since we left Amritsar, and Pete is bitching about his sore feet and about how much he wants a good, cold shower, and I'm getting more and more tired of hearing it when all of a sudden, what should appear in front of us but an ice-cream bar. We're out in the middle of nowhere and there it is, a modern, stainless steel and glass UNICEF Dairy Milk Bar, operated by something called the Punjab Dairy Development. Well, it's as if we're suddenly back in the U.S.A. We sit down at a clean, air-conditioned, Formica-

Waseca, Minnesota: The start of the walk.

Bluffs along the Mississippi.

Madison, Wisconsin.

BRUCE M. FRITZ

David, Pete, and John in Pennsylvania.

WIDE WORLD PHOTOS

42nd Street, New York City.

WIDE WORLD PHOTOS

Starting across Portugal with donkey and cart.

FERNANDO RICARDO

With Princess Grace of Monaco.

Thor Heyerdahl and party inside a restaurant in Italy.

Dave, Willie Make It II, and John in St. Mark's Square, Venice.

CAMERAPHOTO

The Mayor of Trieste signing our scroll.

E. SELHAUS

San Remo, Italy: Having a good laugh.

Outside Ankara, Turkey: Drifter tied to the back of U.S.A.-Turk machine.

Iran: Willie and camel.

Buzkashi: Afghanistan's national sport.

In the Afghan desert.

The morning after at the gorge near Kabul, Afghanistan, where John was killed: Noor Mohammed, John's assassin.

topped counter and have three plates of ice cream apiece, with Cokes: orange ice, chocolate, and banana. As much as we can eat. The guy at the counter doesn't speak any English, so we can't find out about how he happened to be here or if there are other places like this around, but we can't complain.

I remember once in Bulgaria something like this happened. Suddenly, out in the middle of nowhere, an American-type restaurant. It was off the road about half a mile, but we'd taken a chance and followed a sign. It turned out to be a little country inn, like you'd find in Pennsylvania, say, or anyplace in the United States for that matter. It had an outdoor pavilion, and there were waitresses, pretty ones at that, and real food: steaks and draft beer and good tossed salad and french fries. We'd had the place almost all to ourselves, too, just a couple of Russian officers eating over at a table across the room. Then, just as we were leaving, a whole party of officers came in. Apparently, it was some sort of club. Anyway, this ice-cream place was kind of like that.

As we get near New Delhi, there are more and more people and they start to pester us again. The night before we walk into the city is the worst. We're parked in a gas station and between the people and the mosquitoes, we're going crazy. We've just had dinner when Pete catches some guy trying to get off with one of our cooking pots.

"Goddamnit," he yells. "I'm going to kill you bastards." And with that he takes the club and jumps out of the wagon and I'm really worried he's going to do something that will get us into real trouble. What he does turns out to be pretty effective, though, I have to admit. He marches around the wagon swinging the club until there're about three feet between us and them. Then he stamps on their toes with his shoes. That really does it. They set up an awful howl, but they don't do anything, just move back the way they're supposed to. And they stay there, too. Some of them even take off. If one gets any closer, Pete just runs over to him and stamps on his toes. Back he goes, yelling his head off. We should have had helmets with us and GI uniforms and M-1's. That would have done it. We wouldn't have had half the trouble we had if they'd thought we were soldiers. Everyone respects a military man. Even these people.

We spend three weeks in New Delhi and for me the best thing about that is the hot time I have with Lucy May Rivers, the driver UNICEF has provided for us.

"Address yourself to those knockers," whispers Pete to me the day she walks in and introduces herself. Well, it isn't long before she's driving me, all right. She's from Atlanta, Georgia, and is probably the easiest-to-get-along-with woman I've ever met. Right away I have that feeling you get with some people, that you can say anything you want to and they'll understand. They might not like it, but you can say it anyway. You don't have to worry about it, that's the thing. So, she takes us to the apartment the embassy has gotten for us. I stick my head into the bathroom as we go by and just for the hell of it I say, "How's about a shower? You got time for one?" Instead of going off in a huff the way some girls would have, or putting me down, she laughs and says, "What's the matter? You dirty?" But later that same day we're back here without Pete and we take that shower together. Believe me, it's been a long time. The nearest thing to Lucy May I've experienced up to that point is Marie, but that's another story. I was really in love with her. She was a girl I met in France who joined me in Yugoslavia and then later on in Turkey. I might have married her, I think, if things had turned out differently; but for pure, easy fun and such, Lucy May was about as nice as they come.

"Big boy, aren't you?" she says, and laughs, and for the first time in my life I make it with someone standing up in a shower stall. I didn't know it was possible. I'd heard about it. Soldiers on trains. Stuff like that. But I'd never really thought about it as being much fun. "Whoops. There goes the soap," I say, and fumble for it between her legs.

We're both laughing and working up such a good lather between us and around us that we're practically disappearing into it. And then someplace in the middle of it all we get into contact and it's as easy and natural and pleasant as falling off a log and about five thousand times as exciting. When I tell him about it, Pete is jealous and horny as hell, but what the hell can he do?

"You're happily married, remember?" I tell him. "What are you complaining about? I'm the guy in misery."

"Some misery. You want to share some of it?" he says. The funny thing is Lucy May kind of likes him too. I can tell. Probably, if I were so inclined, I might have fixed things up for old Pete as well, though actually I don't think he would have gone along with it even if I had. He's very straitlaced in his own peculiar way, and he'd made some sort of agreement with Nancy before he left, I think. But I give him what I can. I tell him about how I'm doing.

"Cut it out," he says. "Don't torture me."

It's mean, I know. I shouldn't do it. Partly, I think, I'm taking revenge for what John did to me on more than one occasion. Lucy May would have sure loved him all right.

One night we're having dinner at the Weavers'. Don is the Voice of America representative. There's an Indian friend of theirs there, and he gets very excited about how foolish we are not to take better precautions crossing India.

"You're knowledgeable of the dacoits, are you not?" he says.

I nod but Pete says, "No. Who are they?"

"Robbers," the man says. "Completely without scruples and utterly vicious."

I smile. I've heard this stuff before. Pakistan was full of robbers, they told us, but somehow we didn't see any. Sure, there are robbers. I don't doubt it. Everyone in India is a potential robber as far as I can see. But what can you do about it? They have your number or they don't. I stopped worrying about them a long time ago. It's like the traffic.

The man goes on, though. He's very intense. His high, dark forehead, damp with moisture, glistens in the light as he turns to me, then back to Pete.

"These dacoits are professionals," he says. "They would kill you for your mule. You are very fortunate they have not already done so. But the greatest concentration of them lies ahead." I must have let something of what I am thinking show because at this point the man gets quite upset.

"Perhaps you do not believe me," he says. "Perhaps you think I wish to frighten you." I make no reaction. Neither does Pete. The man goes on. 'The fact is my own countrymen are killed by the dacoits every day. Foreigners like yourselves . . . Even in a car you would not be safe. Really, it is appalling to think . . ."

Appalling is it? Don't you think we know about these things? Pete's explaining to him now about our getting shot up in Afghanistan. The man's nodding his head up and down, and smiling.

"Just so," he keeps saying. "Then you must be very careful. Do not leave the Grand Trunk Road. Stop for the night at government rest houses."

"What about the road to Agra?" says Don. "Is that still dangerous? They're planning to take in the Taj Mahal." The Indian just looks up into the air. These Americans, he seems to be saying—even the ones who are supposed to know better.

I am about to explode now, but he answers before I can say anything.

"The very worst place. Please," he says, as if talking to children. "Promise me you will not take the road to Agra."

"If they're such a menace, then why don't you clean them up?" It's Pete talking. I'm afraid he's going to punch the guy. There's a moment of dead silence, and then he goes on. He leans over closer to the Indian. "What's the matter with you people, anyway. No one ever *does* anything. You've got crime and disease and all these stupid . . ." But the Indian is up now and over by the door.

"I have known people personally who have been attacked by dacoits," he says. And then he leaves. We look at each other, but no one speaks. OK, I think. We'll skip the Taj Mahal. It isn't supposed to be any good anyway, except in the moonlight.

Boy, is it hard saying good-bye to Lucy. I've been with her almost every day for nearly a month and now I'll probably never see her again. I don't know which is worse: going through places so fast you don't have a chance to meet anyone or getting to know people and then having to leave them for what is probably forever. I tell Lucy I'll be back to see her. We promise to write. But you know how it is. It's never the same again, even if you do meet up. Partly because of Lucy we've been putting off our departure. Now the Fourth of July's coming up and the embassy's planned a big celebration, so we put it off a little longer and plan for the Sunday after that Wednesday. We've gotten in touch with all the government tourist bureaus along our route and the Indian home minister has given us an official letter and has told the embassy that he will see to it that the police in all the towns we are going through from here to Calcutta are notified that we are coming. We will be well taken care of, he promises. "My country is safe. You have nothing to worry about walking through my country, day or night," he tells us, "as long as you stop for the night in one of our many rest houses." We nod our heads and smile. Sure.

We hate like hell to leave New Delhi. Between here and Calcutta there is nothing really: no real city, no American club, nothing but one grubby Indian village after another. And Calcutta is one thousand miles—forty days at the very least.

CHAPTER

WE'RE TEN MILES PAST THE OUTSKIRTS OF NEW DELHI AND still pushing our way through the crowds. It's only seven in the morning but the temperature is well over one hundred degrees. Whatever happens, we try to keep moving, but sometimes that's impossible. Bicycles! I've never seen so many people riding bicycles. And there're people sitting in the middle of the road or lying down in it, squatting everywhere. It's like some gigantic parade come to rest. "When are we going to get out of here?" I keep thinking.

An hour later the crowd has thinned a little, but still, even in the open, there are so many people per square foot we can hardly see the ground. "Oh, no!" yells Pete, suddenly. I look at where he's pointing, and there, not ten feet from us, is a guy taking a crap. On the ground. Right in among everyone. I turn away in disgust. What a country!

There'll be fewer people for a while, and we think, at last, we're out of it. But then they close in again and they're thick as fleas. The stink is unbelievable—worse now that we've seen what's causing it. We can't stop to rest because the crowds would smother us, so we just keep going, keep trudging down the Grand Trunk Road east toward Calcutta, more as if we were part of some huge caravan than like travelers passing through a foreign country. We'd hoped to make it thirty-three miles to a place called Sikandarabad today, but there's no hope of that. We're not halfway there. Just after dark, though, the people start to thin out and what with that and the relative coolness of the night, things are suddenly much better. We walk till ten o'clock, it's so nice. Then we pull off onto the side of the road.

"If only we could walk at night and rest during the day," Pete says.

"Yeah," I grunt, too tired to talk.

"How far do you think we got today?" Pete says.

"I don't know. About eighteen miles." I lie down in the wagon but Pete just sits there. "Come on," I say. "Let's go to bed." I think Pete's crying. I don't know. He keeps rubbing his feet and sighing these great, big, mule sighs. He's not very happy, that's for sure. It's different for me, I guess. I've been on the walk over three years now. Pete's feet aren't even broken in yet.

Pete pours himself another whiskey. He downs it in one gulp, then staggers off a couple of feet and I can hear him relieving himself on the hard ground. He comes back in and lies down beside me, but then he gets up again and takes his bag outside. "Good," I think, as I go off to sleep. "More room for me." Later on, though, I wake up in the night, and for a second I think Pete has left me. I've been dreaming about Minnesota: sledding down Olsen Hill with the kids. Suddenly I'm in a cold sweat. Where's Pete? I get up and go outside. There he is sitting on a low stone wall next to the road. "Hey, fella!" I say. "Can't sleep?"

He turns. "I've been thinking," he says.

"Yeah?" I want to say more but I can't.

"I'm keeping you back, I'm a drag on you, aren't I?"

"Not really. I mean today it was all those . . ."

"We should go faster, though. As fast as we can."

"Sure," I say. "But what about your feet?"

"To hell with them."

"Just to hell with them?"

"If they get so bad I can't stand it, I'll just get in the wagon, that's all."

I look at him for a minute.

"What I mean is it's not as if it were my walk," says Pete. "It doesn't really matter if I quit."

I hadn't thought of it that way before. "Yeah," I say. "I suppose you're right." But the idea shocks me. Pete could quit the walk anytime, couldn't he? There's nothing really in it for him. Nothing at all if he gets into the wagon. The thought of Pete actually leaving and me walking through this country alone sends uncontrollable shivers through me.

"Come on," I say. "Let's go to bed. We'll talk about it in the morning." But later, lying there next to him, I am unable to sleep. Without Pete, I realize for the first time, this walk would be impossible.

About noon the next day we get to an intersection in the road and I ask a guy in a gas station which way to Calcutta. He points to the right. We go along about a half hour and I begin to have a funny feeling so I ask someone on one of these three-wheeled taxi things which way. It always pays to get two opinions. "Calcutta?" I say, and shrug my shoulders and point. Instead of nodding or shaking his head the way I expect him to, though, the dumb dork looks at me as if I've just insulted him. Then he clenches his fist and shakes it in my face and starts to scream out things. It takes only about a minute before there're people pressing in from all sides. The crazy dork is yelling at the top of his lungs and keeps running over to me as if he's ready to hit me in the face. We're trying to move on but it's impossible. We can't budge in the crowd. Suddenly, the man lunges at my chest, and for a second I think he has a knife. But he doesn't, and when I push him back into the crowd, someone grabs him and starts to lead him off and another man pushes ahead of us yelling something or other that causes the crowd to press back some more and we're able to squeeze through. It all happens so fast, that's what gets me. First of all he's an ordinary dork and then all of a sudden he's out to kill us. Why? Who knows? There're a lot of crazy people in the world, that's my answer to that.

We keep on going until we get to a place called Khurja. There's supposed to be a government rest house here. We've decided to stay in them whenever we can. But how we are going to find it is another problem. I don't feel like asking anybody, that's for sure. Well, Pete finally finds some people who speak a little English. They're a couple of Indian students, it turns out. They come along with us, chattering to Pete the whole way, and pretty soon we're there and they explain to the manager who we are and what we want and everything. They even help us get some feed for Willie. But then they want to stick around and try their English out on us. Well, normally that would have been all right, but after two days of walking the way we've been, all we can think of is washing up as best we can, eating, and going to sleep. So what we do is crawl into the wagon and fake sleep until they leave. It's a dirty trick, I guess. But sometimes there's nothing else we can do.

The government rest houses sure aren't much: a couple of crummy buildings and a well with a pump that doesn't work. But at least there's a compound around them and we're not pestered as much by the dorks as we would be outside. In fact, they'd be fine except that here there's a big family of pestering dorks on the inside serving as caretakers, and if we show them any encouragement they're all over us the whole time, and pretty soon they're driving us just as crazy as a lot more of them would be. The kids come up and smile and say, "Hello, mister." Fine for a while, but they keep it up. So finally we just pretend we don't hear them. It's the only way to make them go away. If we even look over at them, though, they'll be back pestering us some more.

The next day I try something different when the dorks start pestering us. "Hello, mister. Hello, mister. Hello, mister," they say.

"Hello, yourself," I say back.

"Where you going? Where you going? Where you going?"

"Calcutta!"

"How you get there? How you get there? How you get there?"

"Walking."

"How walking? How walking? How walking?"

"Walking with legs," I say. "Walking with legs."

"Legs what doing?" they say.

"Riding bicycle," I say and smile as idiotically as they do. That throws them a little.

"What *you* doing?" I ask them. Pete joins in. "What *you* doing? What *you* doing?"

Most of them start to drift off now.

"Walking?" we say, if they're riding a bicycle. "Riding?" we say if they aren't. If that doesn't get them, we yell, "How riding?" or "How walking?" And then very loud and both together, *"With legs?"* That does it. Off they go. All this takes a lot of energy, though; and we really don't feel like playing games in this heat for very long, so most of the time we just trudge along being pestered and trying to ignore it.

Then there're the trucks. It's a one-lane road along here and every time a truck goes by (there are practically no cars now), we have to get off onto the shoulder, which is all bumpy and full of holes and water from the last rains. After a while there get to be so many trucks that we're doing most of our walking on the shoulder, so I decide we've got to fake them out. "Let them go around," I say. "What do you think they're going to do? Run us down?" Well, they practically do, but not quite. Soon there is a quarter of a mile of trucks lined up behind us. We're feeling pretty good, jumping around and laughing; and then one comes from the opposite direction and for a minute it looks for sure as if he's going to plow right into the whole line of us. At the last minute, of course, he turns, but he sure shakes Willie up some. I think the radiator brushes Willie's nose whiskers.

Then the trucks behind us start to pass. We're walking along the road. There're dorks riding bicycles and walking along next to us and standing all around. And then there's this line of trucks off in the ditch to our left splattering past us. That's the high point. We laugh and jeer at the drivers, and I think, "What if one of them stops and gets out and starts to come for us?" But none of them does. The trouble is that when they've all gone by, we have to go through the whole thing all over again, and half the time Pete or I chicken out when we see a truck barreling along at us, and get off onto the shoulder. So finally it's easier to stay on the shoulder the whole time. But that slows us down so that we're making only about fifteen miles a day, and it's hard as hell on Willie. Already she's beginning to lose weight and develop bad harness sores. You can't walk around the world on the shoulders, that's for sure. So we keep trying to get back on the road. It'll be two lanes for a while and we'll be OK, and then it'll go back to one lane, and it keeps going back and forth like that for about a week, getting hotter and more humid every day, it seems. And then, finally, on Friday, July 13, the monsoon rains hit.

At first we think, good! It'll cool things off. And it does. Pete and I are out in it soaping down and shouting and laughing like two-year-olds and the dorks have all run for cover so we even take off our pants and get a real shower bath. It's the aftereffects that are so terrible. In the first place, as soon as the rain stops, it gets about twice as hot as it was before, right away. It's a regular steam bath. You can hardly breathe. And then the roads: mud! Nothing but mud! Poor old Willie. We have to get behind and push, and pretty soon we're covered with reddish clay and our feet are soaked so that they start giving us real trouble, and Willie's sores are getting worse. What we do finally is get in the wagon and rest when the rains come. Usually they last only fifteen minutes or so, but sometimes they're a lot longer. Then it's like a sheet of water out. You can't walk in it because you can't see any more than you would be able to if you were underneath a waterfall. And, of course, pretty soon everything is wet in the wagon and things start to go moldy and rotten.

Three days later we're parked in a public works compound in a place called Bilheri and we've made twenty-eight miles that day, so we're exhausted, when suddenly Pete says, "Hey, happy birthday!" I've forgotten all about it. I'm thirty-three. It's nine o'clock at night but Pete wants to celebrate.

"Stay right there," he says and walks off down the road. A half hour goes by. Forty-five minutes. I'm about to go to bed without him, when I hear him coming. He's whistling! What the hell's he got to whistle about? And then I see what he's carrying: a six-pack of Cokes! They're even cold. There's dirty laundry spread around all over the place in the wagon so everything stinks, but we sit right there in the middle of it all and down those Cokes: one, two, three. Nothing ever tasted so good. "Where did you get 'em?" I say, belching happily.

"Government secret," says Pete, but he tells me. The guy that runs the gas station. For a little something extra he finds he has some in the back.

"In the middle of 'Nam, you could always get a Coke," says Pete. "If you could pay."

Suddenly I think of Dr. Moede, back at the American Dispensary in Kabul. "What would you like, boy? What can we get for you?"

"Could I have a Coke?" Glenis had to go to the embassy to get it. Old Moede didn't allow Cokes in the place, for some reason. But did that Coke taste good. Even with a hole in my lung! Good old Pete. Six Cokes. Imagine!

The next day we can hardly get Willie to move. She has two really bad sores, and every time it rains it wets the padding we put over them, and then they're slick and raw-looking and there's nothing we can do but stop and rest her until they dry off a little and we can shake some talcum on them and pad them for her again. If only we could change where the harness rubs on her back, but we can't, so there's nothing much we can do but dress her sores the best we can and then make her move out. I pull and Pete beats on her and finally she gets going, but even if it doesn't rain she doesn't go for long, and nothing we do makes her move faster than about half-speed. It's not just the sores either. She's picked up something: worms again, probably, and you can see her ribs she's so thin.

It's about three in the afternoon and Pete's gone in to take a piss at the local gas station and to get some Cokes if he can, and I'm standing outside trying to keep off the dorks and worrying about Willie. Suddenly there's Pete, looking terrible.

"What is it?" I say, feeling something sink inside me.

"I'm pissing blood," he says. "What am I going to do?"

Well, right away I think of all the pills we've got, all the ones Dr. Moede gave us and the others we've picked up along the way. We don't have anything to take if you're pissing blood, though.

"What did you do, hit yourself?" I say.

"No," says Pete. "That's what I don't understand. I haven't done anything."

"You'll be OK," I say. "Let's go." Well, we move out—slow the way we have been. I don't think we've made ten miles yet today. We're going along over this narrow bridge and I can't shake this awful feeling I've had inside me ever since Pete told me about his piss; and then, suddenly, something bangs into the wagon and I look back and there's a damn ox with his horn right through the tailgate and another ox beside him on the ground being dragged along in his harness. The Indian driver of the cart is yelling and carrying on like crazy and if we don't do something fast who knows what'll happen. I throw on our brake and the wheels start skidding, which eases up the strain on Willie, and Pete and I run around to the back of the wagon to see what can be done. Pete jumps up on the step and when I do the same it lowers the tail end enough so the ox has a straight pull. Out comes the horn. I rush back and release the brake and Willie takes off so fast I'm practically knocked down. In a few minutes the crowd's all behind us.

I look back at Pete. "Close call," I say, and grin. He just nods, though. On we go, same as we've been going for so long now I can hardly remember when it started, and then I notice that Willie's practically stopped again. "Come on," I say, but my heart's not in it. Actually I'm looking around for a government rest house so we can stop for the day. We find it, finally, and even though it's only five o'clock we pull in. Pete goes to check out his urine, first thing, and when he gets back I can tell it's still bad, though he doesn't say anything. The dorks in this place are worse than in most. There're about twenty of them and they won't leave us alone no matter what we do, so we get into the wagon as soon as we can and close up the flaps. Anything, even the heat, is better than being pestered by them.

Pete doesn't eat much, doesn't say hardly anything either, and before it's really dark out he's asleep. I open up the back flap just enough to get a little air and watch the wild monkeys playing on the rooftops. There are a couple of palm trees near one of the buildings and the monkeys jump into the tops of them, too, and if the kids aren't around they'll scurry down the trunk and even scamper over close to the wagon. Willie stands over near them, with her head low, and doesn't look up when they go by. "We'll have to rest her," I say to myself. "She can't go on this way. Oh, God, make Pete OK in the morning. Make it all right, whatever it is. We can stay here for a while, if we have to."

There's no doctor, of course. The biggest town near here is Allahabad, a good four days farther. Pete looks so pale and still, lying there on the floor, his face beaded with perspiration. He breathes unevenly. A monkey chatters on the rooftop. Scurrying sounds. More chattering. He'll be all right, I keep saying to myself. But then I imagine him dying. The thought of being alone here without Pete leaves me weak. I tremble violently for a moment and still shaking get up and leave the stifling wagon and walk for a while in the slightly cooler air. I sit on the wheel and watch the monkeys. Then it starts to rain again, so I crawl in beside Pete and try to sleep. I fall asleep and wake and fall asleep again and all throughout the night there is the scurrying and the jabbering of the monkeys and the sound of the rain and every once in a while a feeble bray from Willie. Willie will never make it to Calcutta, I know that. How much farther do we have to go, anyway? Seven hundred miles?

"So what will you do if she dies on you?"

"Walk on with what we can carry," I say.

"You'll never make it."

"Yes, we will," I say.

"Not without Pete."

"Pete's all right. There's nothing wrong with Pete."

"You'll die too. Something will kill you."

"Go to hell!" I say.

The voice laughs. "Go to hell yourself," it says back.

"Go to hell!" I say again, louder this time, and that wakes me. It is pitch-dark and for a second I have no idea at all where I am. I am about to yell out. I feel the cry coming up in my throat, and then I hear Pete moan and I'm back in the wagon, and I lie down again and in a while drift off to sleep.

In the morning Pete is still there beside me. I have my arm around him. My face is buried in his neck. "Wake up!" I say, and shake him.

"Huh?" He turns slightly. His face is soaked in sweat and he looks pale as death in the dim light. "More sleep," he mumbles and turns back again. "OK," I think. "Sleep, then. We'll stay here today. Rest."

It's the morning of the second day in this same compound and we're just finishing breakfast. Pete's looking better, but still not well. "Maybe I should hitch back to New Delhi?" he says, suddenly. I look straight at him and he looks away. "If I left now, I'd be back by Wednesday," he says. Wednesday. That's four days.

"Sure," I say, and grin. He wouldn't leave me here. Not here. He wouldn't do that. "Good idea," I say. "Check it out. I'll catch up on my reading."

Pete looks relieved and actually I feel better, too. Who knows what it is, right? Let him go back, see a good doctor. We'll all take a rest. Maybe I can find a vet to look at Willie. Pete packs a few things together and walks over to the road. I keep looking but he's still standing there an hour later. Finally, a truck stops and I see him get on. "So long," I yell and wave. He leans back on his pack, and then, as the truck goes off, raises his arm and waves back. He's smiling but he doesn't look happy.

"It's OK," I think. "He'll be back soon and whatever it is will be all fixed up." I try to smile but there's nobody to smile at and then all the goddamn dorks come rushing over to me. Where's Pete gone? they seem to be asking me. Am I alone now? I can't explain anything to them, so I just get into the wagon and close it up again, hoping they'll go away. It's stifling hot with everything shut, but I don't dare open up the sides until they've gone, so I sit there trying to read, and feeling more and more depressed every minute. Every once in a while one of the dorks sticks his hand or his head through the canvas and I have to yell at him and gesture him off and it's hours, it seems, before they get tired of pestering and go back to milling around over by the charpoy. I open things up again, but then I'm even lonelier than I was before. There they are twenty feet, thirty feet away, and I have to pretend they're not, that I don't see them. If I relax for a second they all come back again and I have to go inside and close up. I'm in a cage. I'm just like a goddamn monkey in a cage with everyone jeering and spitting and shouting at me, and there's no place that I can go and nothing I can do. Just sit here and wait for Pete to come back. Just sit here and wait, till Wednesday at the earliest. And what if he doesn't come? What if he's really sick and they don't let him come back? What if he doesn't come back at all? Ever?

CHAPTER

10

IT'S SEVEN O'CLOCK THE NEXT MORNING AND I'M GETTING ready to move out. Yesterday was one of the worst days of my life. I don't think I could go through another like it. After Pete left I thought I'd try to find a veterinarian. I hadn't even gotten out of sight when I heard Willie braying, and when I got to her one of the dork kids had already led her around to the back of the compound and had taken her lead rope off. There were two other kids climbing into the wagon. I made a dive for them and then got Willie and tied her to the wagon. I couldn't leave them, so I decided what I'd do was clean out the wagon. Every time I'd get the clothes hung out on the top of the wagon, though, it would rain, it seemed. I tried to grease the wheels, but as soon as I got down under, all the dorks came crowding around and I ,had to get up off my back and chase them away so that they wouldn't steal everything. I didn't get anything done. Spent the whole day standing guard. I thought I could take it, but these bastards were finally getting to me. To hell with everything, I finally decided. Walking, I'd at least be doing something.

I'm walking out of the compound at last, yelling at the dorks to get out of the way. It's worse than I thought trying to do all this by myself, but we're rolling now and I'm not stopping for anybody, believe me. I've got a switch about eight feet long in my left hand, and I pull Willie with my right hand and give her a smack on the rump with the switch every couple of steps. Maybe she thinks Pete's still there, I don't know. Anyway, in spite of her sores, she seems glad to be going, so we start out at a pretty good clip.

There seem to be more people, though, than ever before. They're so thick in places that we just can't get through. If it weren't for the fact that they're all trying to go somewhere themselves, we'd be stopped. As it is we're caught in a tide—a slow, stinking tide of people. Sometimes the tide bursts and we can move on for a while, but then there's a bullock lying in the road, or some other obstruction, and we're pressed in again from all sides—just standing there in the steam and the stink and the noise. And then it rains. And then there's more mud.

A ball of mud splats against my bare arm and in wiping it off I notice that it is not mud at all but cow dung. Furious, I look up and there, across the road, are three kids, practically naked, making shit balls and throwing them at me. A truck goes by between us, spattering everything within twenty feet with red mud, and when it passes a ball of shit hits me on the neck and splatters across my chest and shirt. I drop the lead rope and run after the little bastard, landing a good crack with my switch before he gets away. I run back to Willie before anything happens, but a foot of the switch is broken off, now, and I have to stretch to hit Willie on the ass. That's the end, I think. Little kids throwing shit balls.

But then come the bicycles. They used to drive back and forth, dorks on their way to somewhere or other, but now there's a group that's been trailing me ever since noon. Two or three will ride by together. The first will say something like, "What doing?" The second maybe scratches at the canvas with a stick, trying to poke a hole in it. The third jerks at Willie's lead rope or grabs at my switch or pokes something at me. Whatever it is Willie and I end up either stopped or off on the shoulder. Sometimes Willie spooks and we'll be down in the ditch and it'll take me a half hour to get us out, and all the while there's a crowd of jeering dorks standing around and guys on bicycles wheeling in big circles just waiting for me to get going again so they can start all over. And there seems to be nothing I can do. If I push one of them over as he goes by, ten others come along and they drive me down into the ditch. What I try to do is ignore them.

  Finally the sun starts to go down and the crowd thins out some, and I look around thinking that maybe the bicycle dorks have decided to call it a day. But there they are, charging down the road straight at me. Fifteen of them, at least. They're all carrying sticks and I'm afraid they'll rip the canvas for sure this time or that Willie will spook so badly she'll turn over the wagon; so what I do is lead Willie straight down into the ditch, right into the muddy water and across to the edge of the field beyond. The first one turns off the road and tries to get me anyway, but he slips on the mud and down he goes, and the rest of the line swerves off and stops. I'm standing with my back to the wagon, switch raised, just waiting for one of the bastards to come across the ditch so I can let him have it good. The guy who slipped gets up and I think for a moment that he's coming, but he just glares at me and then pushes his bicycle up onto the road again. The others come up on their bikes and mill about. I'm yelling at them by this time, yelling and beating my switch in the air. "Come on, you douple-dorks," I scream. "Come on, you bastards." I really want them to come, too. I'm bigger than any of them and even if they charge me, I'll take three or four with me, at least. "Come on!" I yell. "You lousy cowards!" All at once they ride off. I'm laughing now, and shouting, and the people standing around must think I'm really nuts, but I don't care. "Douple-dorks!" I shout. "Douple-dorks!" I'm in hysterics. I've never heard anything so funny. All of a sudden I stop. Where's everybody gone? Suddenly exhausted, I sit down on the back step of the wagon. "How far is it to the next government compound?" I wonder. "At least ten miles. The hell with it," I think. "I'll camp right here."

About nine o'clock the moon comes up. It's a still night—scattered clouds. There's a lightness in the air I haven't felt for weeks. If I shut off my mind to where I am, I might almost believe I was back in the States somewhere, back in Minnesota, say, out someplace with the survey crew. Once in a while we'd get caught late on a job and someone would have to spend the night with the stuff; sleep in the truck. A lot of times it would be me. I was the boss, but I'd do it anyway, because I liked it, liked to be alone out where it was quiet, glad to have an excuse not to go home for the night, particularly that last summer when Jan and I were barely speaking.

There's a little breeze and no mosquitoes and I'm sitting on the steps with the wagon open and everything, looking out at the moon that is full and kind of pink and slowly rising. I think of how many times I used to do this sort of thing when I was a kid. Mom would drop me off in the car and for three days I'd be in the woods, on my own, hunting.

"It's supposed to taste like chicken," I said.
"Well, it sure doesn't smell like it," John complained.
"Believe I'll pass on the raccoon, son"—Dad. That funny smile of his.
"I cooked it," said Mom. "But I'm not saying I'll eat it."

Mostly I shot rabbits and squirrel. A lot of times I just walked, though. I'd leave the gun and my other stuff and spend the day tramping in a huge circle through woods and across fields. I'd cover twenty, thirty miles, and be ready to drop when I got back. Mom would pick me up at the time we'd arranged and I'd be bubbling over with things to tell her all the way home. It was sort of a joke between us.

"Time for a trip," she'd say. Sometimes she'd drive me out there and dump me when I only half-wanted to go. She was always right, though. It never failed but I'd come back in a good mood.

I'm sitting bolt upright, the muscles in my shoulder quivering from the tension, holding the knives out in front of me, ready to stab them down through the hand that is slowly creeping toward my left foot. Behind the hand there is a foot-long slit in the canvas. It was the sound of the knife that woke me. I'd reached up to the roof of the wagon and pulled the knives out of their sheaths before I was fully awake, I think. First the faintest whisper of slit canvas, then the hand exploring the inside of the wagon. The hand stops. I wait, tense as a wire. There're probably at least six of them, I think. In another minute I will hear the scratch of cut canvas, perhaps behind me, and there will be another hand inside, searching me out. This is what they do. They are masters of silence, of knives, of the night. Usually their victims are found in their beds, their throats cut from ear to ear. At a touch, their heads flop over, like box lids.

The hand moves, touches my bare foot, and instantly recoils out through the slit in the canvas. For five, for ten minutes maybe, there is nothing, not a sound; then a hand slips in through the side flap at the front of the wagon and moves slowly, first to the left, then to the right. I curse the hand just beyond the reach of my knives. I ready myself to lunge forward, but just then the hand leaves. Nothing. Silence. For what seems forever. Then I hear it, the canvas being slit behind me. I pivot silently, ready; and when the hand is almost to my side, I bring my knife down onto it with one sweeping motion and pin it to the box. There is a sharp cry, a quick movement, and I see that the hand has been torn backward through the still-standing knife. Sounds of running feet. I grab the club and push aside the back flap. Crossing the ditch, running toward the road, are three figures, a man and two boys it looks like. One of them is shrieking, doubled up, holding himself. Outside, I almost twist my ankle stepping on a pot. There are canned goods strewn about. A plastic water jug bobs in the ditch. Not the dacoits after all. I sigh with relief. Only robbers.

But where is Willie? I run over to where I tied her and she is not there. "Willie! Willie!" I shout, and stumble around in the field looking for what I know, suddenly, will be her dead body. "Willie! Where are you, Willie?" I scream it at the top of my lungs, then trip over something and fall. What will I do? I cannot imagine myself walking through this country with just a pack—no wagon to get into. I jump up and start to run, shouting Willie's name. The field is soaked and squashy and in crossing it I fall several times. There is a water-filled ditch at the end, and something scurries off under my feet as I plunge into it. In the next field there are forms of animals—bullocks, water buffalo. A loud snort. Something lumbers toward me. I run back and skirt the field, searching always in the darkness for Willie's form. "Willie!" I shout. "Willie!" A bray. Is it a bray? I dash in among the animals again, but I can see nothing that looks like Willie. At the end of the field, exhausted, I sink to my knees on the slippery side of the ditch, panting, my side aching. I am empty and dead now inside. Willie could be anywhere—she is probably dead. Anyway, there is no hope of finding her now. I turn and walk back along the top of the ditch. Rats. The scurrying of rats in the slime. Halfway across the field, back toward the road, I hear the bray again, then another. I look up. There she is: Willie, my Willie, quietly munching grass in front of me. I hug her around the neck. My Willie. My lovely Willie. Oh, Willie. Willie. Willie. We walk back together across the field to where the wagon is outlined against the sky in the pre-dawn light. "You could have warned me," I say, one hand on her cut lead rope. "You could have brayed." She brays now, as if in answer. When we get to the wagon—it's a good quarter of a mile—I sprinkle some powder on her sores, arrange the piece of flannel blanket that serves as a dressing, and hitch her up. As early as it is, the road's already crowded by the time we pull out. The sun rises like a ball of fire and we walk once more into the furnace of the Indian day. It's still another whole day till I can even begin to expect to see Pete.

The bicycle dorks, the douple-dorks, are worse today than ever. And there are about twice as many of them. The canvas is ripped now from the robbers and before long they've slashed it loose along the whole left side. About three o'clock it starts to pour, and I have a welcome rest inside the wagon, but as soon as the rain stops they're back in force again. This time one of them goes off with a water jug and at the same moment another grabs the half-full bag of grain—all that is left of Willie's feed. I drop Willie's lead rope and chase after them, but the crowd stops me and forces me back. Suddenly, I go to pieces. I will not let them escape. I hit out blindly. I feel the crunch of bone on bone, the taste of blood; my eyes are full of sweat. I bend over and cover my head. Blinding dust. Feet kicking at my ribs, a blow just above the kidney. I am down and helpless. "How silly. How stupid it all is," I think.

From miles away, it seems, the sound of English. Someone is standing over me, offering his hand. He is speaking to me in perfect English—an Indian, but well dressed, educated. "Come, come, sir," he says. "Get up, won't you?" Then he is lifting me to my feet and I stumble after him and in a while we enter a quiet place and I sit down and am brought some tea; and the man is telling me to keep calm, that everything is all right, that his servants are guarding the wagon, and that he will lead me himself to a government rest house where I can spend the night in peace.

Ten minutes later I'm still trembling. We're walking now. "Make way! Make way!" he keeps saying, along with some other things in their language. And the people move back for us. I can't look at them. I can't raise my eyes to look at them. I stare at the ground in front of us, at the road itself, sideways at Willie, and I see their feet and legs and an occasional glimpse of a person sitting or squatting by the side of the road. Even that brief glance is too much, though. I'm shuddering all the time now, all over—like a horse trying to rid himself of flies. It's not that I'm afraid of them, exactly. It's more that I'm afraid of what I might do. Last night. Pinning that dork's hand to the box. Hearing him scream. Watching the knife rip his hand in two as he pulled his arm back out of the wagon.

It made me feel so good. But I cannot let that take over. Because I have to finish the walk. That thought alone is what keeps me from going berserk. If I give in, if I cut loose and kill someone, the walk will be over. It almost *was* over back there. It would have been, for sure, if that Indian hadn't come to my rescue. So I don't let myself look up. I think about the walk. One step at a time, I say to myself. Just keep walking and sooner or later you'll have walked out of all this.

"Here we are, sir," says the Indian with the English voice. We're entering a compound. I see some grass so I drop the lead rope. Cautiously, I look up. My Indian friend is talking to a man in front of the door to a government rest house. I see the heads of two children peeking out from behind the man. I turn and walk over to a basin and try the hand pump. Miraculously, water gushes forth. Not since New Delhi have we found a pump that was in working order. The basin is clean, too, and I fill it and plunge my face right in. It's so clean, so cool. Without thinking I drink it. I pull off my shirt and start to bathe my shoulders, my chest. I pour water all over me. If only the basin were bigger . . .

I spend the night here and the next day my Indian friend brings in a vet who gives Willie a shot and puts some ointment on her sores. "Rest," he says. "More resting," and shakes his head up and down emphatically.

"Sure," I say, and even manage a smile. I spend my time

doing laundry and cleaning up the wagon. Willie seems happy to munch away at the grass.

The Indian family that is taking care of the place hardly bothers me at all. The man speaks a little English, but I steer clear of him anyway. There's the continual sound of people jabbering and shouting outside the walls; and of course there are children at the gate all day, watching me and begging. I need only keep my eyes away from the gate, though, so that's all right. My wagon protects me. The constant din; the noises, the smells of India; these rise and crest and rush upon me like waves. I do not look at the man when we speak; still, I feel safe here, as in the quiet center of a pool. "I will wait here until Pete comes," I say to myself.

All day long I vaguely expect him. Any variation in the constant din outside, any movement at the gate, at these my heart jumps in my chest and my throat goes dry, and I look up quickly, expecting to see Pete there. Yet all the time I say to myself, "How foolish. Today is only Wednesday. The earliest he said he would be back. He may easily be gone a week." A week. He may be gone much longer than that. What if he has something serious? They could keep him for weeks, for months. "Clean the wagon! Don't think of such things," I say to myself. "You're safe. You have nothing to fear. You are lucky to be where you are instead of in a muddy ditch or beaten to death in the street, dust and filth covering you; you lying helpless, bleeding; in pain, dying, while the dorks jeer and shout and kick you with the horny soles of their feet. Be grateful for where you are and to the person who brought you here and to the family that stands guard over you."

Where has he gone? I wonder, my Indian friend? Couldn't he have stayed till Pete got here, my friend who speaks such good English? Where are you, Pete? Are you alive? Did you get to New Delhi or did the dacoits kill you? I see Pete's face as it was when he drove off—the half-smile, the worried look. The thought of hitching rides to New Delhi makes me tremble all over and I climb into the wagon, sit on the box, and, hunched over, tell myself to stop it. "Control yourself! Pull yourself together!" Blood. I taste blood. My lower lip is bleeding. I pinch hard into the under skin of my upper left arm and cry out with the pain. "There! Think of that! Think of that night in the Kabul Gorge! You lived through that, didn't you? Then you can stand this. This is nothing, only in your mind." And then I start to cry. Tears cascade down my cheeks. John is dead and so is Pete and soon I will die. We will all be dead and it will all be for nothing. I let my head drop onto my crossed arms and allow my shoulders to rise and fall, in great sobs. For John I cry, for Pete who has a family, who should be home, who came only to help me.

Afterward, I feel better, though weak. I go to bed early but toss and turn in the heat. All night there are bicycles circling me in my dreams. I see the vacant eyes of a starving child—hard belly protruding above legs like sticks. I turn away and there is Pete's face, ashen white and sweating. He is trying to say something but I cannot hear him. "What is it?" I yell, and wake myself up. I look around me in the dim light of pre-dawn. Suddenly I realize that Pete must have passed me in the night. Right now he's probably looking for me on the road ahead. I must leave. He'll come back when he doesn't find me, and I must be there on the road when he passes. Why haven't I realized this before? How could he know I was here? Fool! Dork! Feverishly, I pack, hitching up Willie in record time. The street outside the rest house is relatively empty. I look around to see if by any chance Pete is there, as Willie and I move down the road into a sky that is as red as blood. An old man in rags, sleeping beside the road, gets up as we approach and comes bobbling over to me, his hand outstretched. I lower my head and pass on.

All day I walk and never look up. Mostly I feel nothing, nothing at all. Between each step there is nothing and each step is nothing, too. So there is just the walking, the physical walking. I keep my eyes on Willie's feet. I match my steps to hers. When we stop for a half hour and I sit in the wagon during a downpour, I am surprised to see that my shirt is torn and covered with blood, my arms and legs smeared with mud and cow dung, and my back covered with welts, as if someone has beaten me there with sticks. I have no memory of anything except Willie's hooves before me. Clop. Clop. Clop. One step at a time. For each of us. Together.

It is late in the afternoon, the sun is nearly down, and something unpleasant is happening to me. They are trying to stop the wagon. Someone keeps pulling Willie from the other side, keeps trying to make her stop. There are voices shouting at me. I tighten my grip on the switch. Let them pull Willie's head one more time. Let me hear those voices once again, and I will look up for just long enough to strike them down. All right! I lift my eyes and there before me is a figure. I raise my switch, but before I can bring it down, a hand reaches up and grabs it from me. I am helpless now. I start to fall. Two arms grab me. I am held in their embrace.

"It's me!" says a voice.

"Me?" I say. And then I see him. "Pete? Is it really you?" I weep, floods of joy, and Pete holds me. Slowly, he leads me off to the side of the road and we sit down. Still, I cannot look around me. Only at his face. I reach up and touch it. Yes, it is really Pete.

"It's all right," Pete keeps saying. "I'm back."

"Back?" I answer. "Back?" But I smile. I'm so tired. I walk over to the wagon. I start to crawl in. Pete can take over now. I will lie down. But Pete won't let me. He pulls me out and we start walking again.

"Only another mile," he says. We walk together in the front, his arm around me for support.

It looks like all the other compounds except that there's a church inside of it. I look up at Pete. Where are we? A man in a priest's black robe walks over to us as we come in. "Father Soza," he says, smiling. "Welcome." He takes us into a room that feels almost cool and is very clean. I sink into a large, stuffed chair, but straighten up again instantly with a sharp cry of pain.

"We must get him to bed," Father Soza says, and I see Pete nodding, his face drawn and serious. Father Soza brings me a glass and says, "Drink." I drink the cold water and look over at Pete who sits across from me.

"What's the matter?" I say. "Aren't you glad to see me?"

Behind Pete is a mirror. A creature is looking at me that I have seen before but that I do not quite recognize. Except for his clothes he looks like any of a thousand beggars along the road. Yet he is somehow familiar. I turn, painfully, in my chair, to see him face-to-face, but there is no one there. The hollow-eyed, filthy specter is me. The faint puff of a cry escapes from my lips as I turn my eyes back to where Pete is sitting. But he is not there. He is kneeling next to me; my head rests upon his chest. I feel myself being lifted into the air, and then I am dimly aware of the voice of Father Soza saying, "Here. Put him in here, on my bed."

CHAPTER

We stay a week with Father Soza and when we leave I'm pretty well recovered. Willie is much better, too, except that it has been raining almost steadily since we arrived so her harness sores have not had a chance to heal. In spite of that, we step out early on the morning of August 2. Willie has been shod and her worms seem to be gone. My back still hurts me some, but the skies are clear and Pete's there behind me and in good shape again. (All it was was a kidney infection.) Only a little more than six hundred miles, which seems like practically nothing, and then India will be behind us. We should be able to do that in three weeks, I figure. That means Singapore by the end of August and Australia some time in September. Two thousand miles across Australia—figure one hundred days—hell, we should be at Pete's by Christmas.

Father Soza must be one of the kindest men I've ever met. He has nothing. Has to haul water like one of the natives, lives on chapati and handouts from his parishioners. But not only doesn't he complain, he actually seems to enjoy his work and is pathetically grateful when we include him in our beef stew or corned beef hash meals. Among other things he runs a leper program. He comes from Portugal and has been out here three years. When we leave there are tears in his eyes. He gives us the names of other priests we can stay with on our way and urges us for the hundredth time to be careful of the dacoits. He will pray for us, he says, "Always. Seven times a day." We thank him and wave and he embraces each of us, and finally we're off: pushing through the crowds, shouting at the people to get them out of our way, making the trucks go around, giving Willie a touch on the ass every once in a while to keep her pace up, our shirts already soaked through in the morning heat.

"Feels good to be walking again," yells Pete, and I look back and he's smiling. I nod my head and smile back. We even sing some.

We haven't gone five miles, though, when Willie starts to slow down. At first it isn't much, but then she gets slower and slower and pretty soon she's going only about a mile and a half an hour, and the worst thing is that nothing we do speeds her up. Pete beats her on her ass till his arm aches, and I pull at her head at practically every step, but nothing seems to affect her. It's as if she'd thought from the way we were behaving that the end was just around the corner and that now that she's decided that the corner's well past, she's going to do the rest at *her* speed, like it or not.

"Goddamn mule!" Pete yells, and beats her with his other arm. "What's the matter, Willie?" I keep thinking. "You've made it practically seven thousand miles. Can't you make it another six hundred?" But Willie is oblivious to both of us, apparently. She doesn't quite stop, but at this speed we don't push our way through anything. We're the ones that have to do the moving.

Late that afternoon we practically come to a stop when all the traffic has to detour off to the left because there's a bridge out ahead of us. It's the Ganges, the holy Ganges, in flood because of the rains. What a mess: filthy, reddish-brown water, miles across it seems. The bloated body of a cow is caught up against one of the pontoons. Nearby, a man and his family are standing waist-deep in the water, the mother holding a small girl in her arms. As we skirt the river we see that thousands of families are camped along the swollen banks. There are foul-smelling bonfires everywhere, processions of people, music that sounds like a hundred cats wailing. They're burning their dead, we find out.

It rains off and on during the afternoon, which slows us down even more, and by dark we're out in the middle of nowhere. It's too far to Saidabad, which is the only place where there might be a government rest house, so we pull the wagon off the side of the road and spend the night in the open. It's kind of peaceful, actually, except that I'm awake half the night listening for dacoits. When it rains, I sleep. But the silence wakes me with a start. My eyes grow accustomed to the darkness and I watch for slits in the canvas, hands, the glint of steel. Pete sleeps on.

The next day Willie's pace is just the same. Now she barely lifts her feet off the ground; and each day that follows, though her speed does not change, her head gets lower and lower. If there are potholes, or we have to walk on the shoulder for a while, Pete has to get behind the wagon and push. Ten days go by like this. We make about one hundred and fifty miles. By then all of Willie's ribs are showing.

"Goddamn mule," Pete says, and beats her with his switch.

"Come on, Willie," I shout, and pull at her head rope. A dull fear grows in the back of my mind that one way or another we're not going to make it to Calcutta after all. We've spent most of the nights for the last two weeks pulled off on the side of the road. Courting fate. And what chance is there now that Willie will make it? She could drop in her tracks at any moment. And where would we find another mule? We haven't seen one since western Turkey. We plod on, mostly not thinking of anything, in and out of the wagon when it rains.

"*You* go back and push," screams Pete. "It's your walk. Not mine." One whole day I let him lead. He doesn't like that any better, though, and we make even worse progress. I try to reason with him. "Just don't talk to me, will ya?"

he says. "Just close your goddamn mouth." So I do what he says, because I know I have to keep him with me no matter what. Just the thought of him leaving me here sends shivers through my whole body. We trudge along hardly looking to the right or left. The dorks can't get to us the way they did before. We hardly notice them. But if I were by myself! We move so slowly that sometimes it seems as if we aren't making any progress at all, that the Grand Trunk Road is, in fact, a treadmill. But we keep on, the three of us, in silence now.

"Oh, I say, you must be the American brothers." The person speaking is a large woman with a very red complexion and masses of untidy gray hair that billow out from underneath her white pith helmet. She's wearing khaki shorts, as is her smaller, trimmer companion, and is more or less blocking our way as we try to push through the packed streets of a place called Dehri-on-Son.

"Come round for a cup of tea, won't you?" she adds, when I nod.

"It's quite near here, actually," says the other lady, and I manage a smile and the first one says, "Just in the next block."

They're English missionaries, it turns out, been here for thirty years. They end up by inviting us to spend the night. "We'd like to," I say, "but there's no compound and we can't leave the wagon unguarded."

"Don't worry about that," says the big one. "Siegfried will take care of things."

"Who's Siegfried?" says Pete. The big one gives a sharp whistle and a huge German shepherd comes bounding out from behind the house.

"That's Siegfried," says the smaller one. "He can't abide the natives, nor they him, if the truth be told."

"Siegfried will watch the wagon, won't you, good boy?" says the big one, bending down and letting the dog lick her face. Already the area is clear of natives. Siegfried is apparently well known. Siegfried's tail bangs back and forth with happiness at the thought of such responsibility.

We unhitch Willie, tie her behind the wagon, and follow the two ladies into their house.

Tea and biscuits are followed almost immediately by dinner: leg of lamb, boiled potatoes, and fresh peas, with deep-dish apple pie *à la mode* for dessert. The whole thing is pretty hard to believe, what with all the English furniture, and the ceiling fans in the rooms, which make things almost cool inside, and the two ladies dressed in what look like ball gowns when they come sweeping in for dinner.

"No, David, you dip your fingers in that. It's not for drinking," says Miss Yardley. She's the big one. So I learn about finger bowls.

"Of course it hasn't been at all the same since independence," says Miss Offutt, handing Pete a tiny cup of coffee.

"Worst thing that ever happened," says Miss Yardley.

"Quite," responds the other.

Pete and I are fairly speechless. They've read about our walk but they don't ask any questions, whether out of politeness or disinterest, I'm not sure. Mostly they lecture us about the evils of misrule in independent India, about the difficulties in obtaining supplies—about the general deterioration of life in the provinces. They won't let us go in the morning till we've had what they call a "proper" breakfast: oatmeal, sausages, eggs, toast with marmalade, and tea. What we'd really like is Siegfried, but his type is hard to come by, and as for Siegfried himself, the ladies have already told us they wouldn't feel safe without him.

"We couldn't go to market," says Miss Yardley.

"Certainly not," echoes Miss Offutt.

"Cheery 'bye," they sing out, waving energetically as we leave and plunge once again into the human sewer that is just outside their doors.

Siegfried makes me think of Drifter, Drifter the First, that is. He used to ride up on Willie's back; that's how it happened we let the little mutt stay; he and Willie were such good friends. From Yugoslavia to Turkey you might say that Drifter was just a good pet, but from there on he was a real watchdog. Right away, too, he hated the Turks. He'd bark and snap at them if they came too close, and warn us in the night if there was anyone around. He had a mean growl for such a little dog. John adored him; never seemed to get tired of playing with him. All day long it was "throw the stick" or "sit up" or "roll over" or some damn thing, and at mealtimes he practically ate out of John's plate. John built a little box under the wagon for Drifter when we got the U.S.A.-Turk Machine in Istanbul, and that's where Drifter rode most of the time then. One whistle from either of us and he'd be out, though. "Sic 'em, boy!" Up would go his hair, his lip would curl, showing his teeth, and all in all, he'd look as fierce as any Siegfried. It wasn't just looks, either. He could whip most dogs twice his size.

And then one freezing day, in the mountains of western Turkey, he ran over a little hill beside the road and never came back. We heard this terrible barking. John took right off with the shotgun, but the dogs were out of range by the time we got there and Drifter was ripped to pieces. They were huge mongrels, three of them, the kind the sheepherders keep to kill wolves; great jagged spikes on their collars to protect their throats. It was a mistake, a piece of bad luck, that's all. Drifter just happened to bump into them. It wasn't anybody's fault. But we found out later that it was just as well John hadn't gotten there in time to use his gun. "You shoot a Turk's dog, make sure you shoot him, too," a Turkish officer told us. Not that that would have bothered John at the time.

If anything, Willie moves even slower now than she did before we stopped at Dehri-on-Son. Her sores are no worse. For some reason they stay just about the same. But she's developed a wheeze, and she's so weak one of us has to push the wagon nearly all the time, now. Partly it's the country we're going through. It's hilly and even hotter and muggier than before. There's a lot more vegetation. We hear the shriek of parrots and other birds and there're monkeys all around. We'll stop for a rest and they'll be on the road begging for handouts or jumping into the wagon to find things for themselves. As the days go by the jungle gets thicker and thicker. Every morning it's like walking into a steam bath. Willie keeps stopping all the time. She'll just stand there for maybe twenty minutes. And she doesn't pay any attention at all to what we do. It rains a lot, too, which makes it all even worse. We're making only six, eight miles a day.

"There's only one thing to do," says Pete one night. "Leave her here. Walk on without her." I look over to where he's sitting, wiping up the remains of his canned beans with a piece of bread.

"It's only a little over one hundred miles, now," I say.

"That's just it," says Pete. "We can walk it on our own."

"What would we do with Willie?" I say.

"Leave her with some farmer. We can come back later on or we can just leave her. What's the difference where she goes? One of these farmers'd probably love . . ."

But I'm not listening to him anymore. I'm thinking of that night when I thought the dacoits had killed her. I'm thinking of all those times she ran off from John and me, trying to get back to those police barracks of hers in Portugal. Spain. France. On the beach. Taking a swim. We'd tied her to a log. When we came out she was gone. Five miles back down the road. She could have killed herself hobbled like that, caused an accident. Just before we got to Italy. Broke that rope for the second time. It took us a whole day to find her. I'm thinking of all the miles she's walked, of all the times she's fallen to her knees trying to pull the damn wagon up some steep hill, of all the shoes she's worn out, of how strong she used to be, and how weak and pathetic-looking she is now. I'm thinking of how she'd stick her head through the front flap of the canvas and we'd feed her a piece of bread or let her lick the pot. The first time she did that was in the mountains of eastern Turkey. It was freezing out and there was no protection for her except next to the wagon or behind some trees: those early mornings when we'd come out of the wagon and there'd be a foot of snow on the ground and Willie would be just standing there waiting for us to hitch her up. I'm thinking of that night in the Kabul Gorge when I told her to go, and she just stood there and looked at me and then let out those little, low brays. I'm thinking of how she whinnied and pranced when we got back to Kabul; of how fat and sassy she'd gotten and how strong she was and how we could never have made it across Europe without her, not to mention all this far. I knew that once we got to Calcutta we'd be leaving her, but I wanted to treat her right. I wanted to spend a little time and find her a good place, something maybe half as good as what we'd taken her from; not leave her here with some dork where she'd either die right away, because he'd spend her feed money on himself, or she'd recover just enough to slave for him for the rest of her life. I know it sounds silly, but more than anything else, I want her to finish the walk through India with us. I want her to make it to Calcutta, too.

"No!" I say.

"She's going to die on us before we get there."

"No, she's not."

"Why not shoot her? Put her out of her misery?"

"We're going to walk into Calcutta with Willie pulling the wagon if we have to push her the whole goddamn way," I shout. "We're going to get Willie to Calcutta if we have to put her in the wagon and pull it ourselves. We're going to . . ."

"All right, all right," says Pete. "Take it easy." In a minute he sidles over to the front of the wagon and gets out. I'm still glaring at him when he comes back in again later on. "You want to walk on ahead, go on," I say. And I mean it. I almost want him to.

"Forget it," he says. "Just forget I ever brought it up." So on we go—into greater and greater heat; and, after the jungle, more and more people.

It's about noon, three days later. Something's down in the road ahead of us and we've come to a complete stop. We're trying to find out what it is and at the same time keep the dorks away from the wagon. Then, suddenly, the line starts up again. I pull at Willie, but instead of moving, she just stands there. Pete hits her a good crack, but she doesn't even flinch. We try everything but nothing works. She won't move. We try to get Willie off on the shoulder at least, but we can't make her take so much as a step. We sit there, like a rock in the middle of a stream, forcing everything to bulge out around us.

"What are we going to do? Sit here all day?" Pete says. Apparently so. Finally, just before dark, Willie pulls the wagon off onto the shoulder and stops again. So that's where we spend the night.

"I'm not sitting around here all day tomorrow," says Pete.

"Me neither," I say. But what we'll do if she refuses to go, I haven't thought out. When tomorrow comes and I pull at Willie's lead rope, she responds. "Thank God!" is all I can think. All day we crawl along. But we crawl along toward Calcutta. We are on the outskirts already, though, still thirty miles from our destination: the American consulate. "Come on, Willie," I keep saying to myself. "Come on!" I think, as I try to will strength into those feeble legs. For as long as I can, I push behind the wagon. Then Pete, reluctantly, changes places with me. Our excitement is building, though. By the time we stop for the next night, we are within five miles.

The next morning, crossing a bridge that spans the Ganges, a bus crashes into the rear of the wagon and Willie goes down. She lies there in her traces, wheezing heavily. I think, "This is it!" But no, with great effort, she struggles to her feet, her worn-out shoes slipping on the pavement. We cross the bridge and I am about to stop and rest her when I see something in her eyes that tells me not to. So on we go.

Calcutta is the worst place yet in terms of filth and stench and press of people, but today none of that matters. Calcutta is not even fully real. All that exists is the moment in that day when the walk will be over. The sense that, except for the mere walking, it *is* over gets more and more certain with each hour until, finally, by late afternoon I am walking in a trance. Balloonlike we drift toward the American consulate. We float through the streets, the three of us and our weightless wagon. It's as if the final picture has already been taken and I'm looking at it in the paper. Nothing, nothing in this world could possibly stop us now.

"Who goes there?"

"The Kunst brothers," I yell out. A cheer. The marine guard is saluting. Pete and I salute back. It's 8 P.M., the evening of August 27. The walk through India is over.

CHAPTER 12

THE PARTY'S BEING GIVEN BY TWO GIRLS, AND WE'VE BEEN invited through a woman we know at the UNICEF office in Perth who lives next door to them. We're celebrities. They want to meet us. Hot damn! We walk into the place and a cute-looking chick comes over and starts bubbling about how great it is we've come and her friend Jill does the same and I can see people whispering and pointing in our direction. Pete's enjoying every minute of it, but I'm just looking around and feeling sort of detached. Parties, parties, parties. Ever since we left Calcutta it's been one party after another, it seems. What's our new mule going to be like? That's what I want to know. Never pulled a wagon. That sounds bad. Almost two months of looking and all we can find is a pack mule. If only we could have brought Willie. But the Aussies don't let animals into their country, and even if they did it would have cost us an arm and a leg to ship Willie all that way.

"Oh, but wasn't it just terribly dangerous walking through all those places?" a girl, whose name I haven't caught, is saying. "Why, I remember when I visited Rome how terrified I was by all those men. I simply couldn't believe it. One of them actually put his hand out and patted me, here." She places a small, fat hand on a large, fat breast. "Imagine!

Of course, if I'd had you with me, he wouldn't have dared." She pauses slightly, then goes on. "But I was alone. I felt so defenseless, you can't imagine."

"Excuse me," I say and move off in the direction of the bar.

"Here's to ya, mate!" Red face, frog eyes, sandy-looking, handlebar moustache: bank teller trying to look the explorer.

"Cheers," I say, lifting my glass, and push past him toward the porch. Once outside, I loosen my tie, take off my coat, and slump into a seat. Parties. These Aussies, all they know how to do is drink beer. And go on blanket parties. Pete and I on the beach with those two girls from the hotel. OK, so Katie's not so bad, but the one I had, well, you have to be pretty hard up, that's all. And what can you do under a blanket, with all those people spread out around you? Plenty, Pete says, which is right if you're a pinch-and-giggle Romeo the way he is. I swear, he could spend half his life under a blanket with a girl and never even find out whether she had a bra on.

My girl's got a nice body, but she's nervous about using it. She kept wanting to talk, instead. Finally, I rolled over and went to sleep, and when I woke up she was gone. "Dummy!" said Pete. But what did I care? If talking was what I wanted I could talk to Pete, couldn't I? There are plenty of fish in the sea. Anyway, I've been spoiled, most recently in Singapore by Nayda. She could talk. Her English was pretty good, as a matter of fact. But what she would rather do was make love. Cute little Malaysian girl. Worked for General Electric—IBM—one of those big American companies. Not sure what she did. Receptionist, maybe. Pretty enough. Loved Americans. Told me that right away. Loved a good time, going out: dinner, dancing. After India she was just what I needed, too. I could have stayed in Singapore another month easily. But here it was practically Christmas. The Australian parties couldn't match what we'd had in

Singapore, not by a long shot; but even if they had, I think I'd have had enough by now.

    I finish my drink and go back into the room to get Pete. We haven't been here long, but this is our second party for tonight. It's twelve-fifteen. The hotel's a twenty-mile drive, and we're supposed to check out that mule at the police barracks early in the morning. Pete's dancing, though. I'll never get him out of here now. He's beaming. From the looks of things he's lucked out. I stand there, my eyes going over the bare back of the girl he's with—smooth, silky-looking in the dim light. They turn slightly, and I follow the curve of her dress to where it goes up behind her neck. She's a beauty, all right. I catch Pete's eye. He moves back a little, bends his head, and the girl looks over at me and smiles. I feel my face turning red. I scowl in their direction, raise my empty glass, and make a slight bow. "What an ass," I say to myself immediately. Then, not knowing what else to do, I just stand there and watch them dance. She's a beautiful dancer. She makes even Pete look good. Moves her hips, her whole body flows. I start forward, to cut in on Pete, then stop. I turn and fix myself another drink instead. What would I say to her? Sipping the Scotch and water I stand there at the edge of the dance floor, staring at them like a dummy. Somebody says something to me but I pretend I don't hear and the person goes away. She smiles at me over Pete's shoulder as they turn in front of me and I feel my face smiling back. She pouts her lower lip and I smile again because I know what she's thinking. She leans her head back and laughs and I know what it is she's laughing at. I know her, that's all. It's as if I've known her for ages.

    Reaching behind me to put my glass down, I hear it tinkle as it drops to the floor. What my legs want to do is sit down, but I force them to walk over to where Pete is. I put my hand on his shoulder, he turns and goes away, and I have her in my arms.

    "You're David," she says. "You don't have to say a thing.

Pete's already told me all about you." She has a soft, ripply sort of voice. More English than Australian. I smile down at her and nod my head. "Pete pointed you out. 'He's the tall, skinny guy staring at you over there,' he said." And she laughs. Her laugh makes the ripples in her voice into little waves. I nod again, smiling even more foolishly. "Do you always stare at girls that way?" she says.

"Not always," I say, feeling my face turning red. I pull her closer, but not so close that I lose her face.

"You're staring now," she says. I pull her in to me and we dance.

I don't know how long we dance, but too soon someone cuts in on us and I'm back at the edge of the floor watching her. It's hard to believe now I ever touched her. Instead of getting a drink, I bump my way back through the couple on the dance floor and cut in on her again. It's like coming in out of the cold.

Suddenly everybody's leaving. It's two o'clock; Pete and I are about the only ones around. Jenny, that's her name, walks over to her car. I'm right behind her. Pete's there too. She opens the door and turns, smiling. . . .

"Sure, we'd love . . ." I manage to beat Pete onto the front seat.

"Can you follow us?" she calls out. I slam the door, lean back, and turn toward her, but instead of saying anything I just follow her hand as it reaches over my knee to turn on the ignition. She looks at me and I grin like an idiot.

"Starts right up, doesn't it?" I say. Then I laugh, much too loudly. I feel my face turning red again.

"You'll like my house, I think," she says matter-of-factly. "It's sort of different."

Pete keeps right on our tail, the bastard. He's at the door before we are. She shows us in, turns on the lights. "Beautiful," he keeps saying. "What a place! The kind of house you can imagine an artist living in, don't you think, Dave?" Get lost! That's what I think, blowhard.

"Yeah," I say. We follow Jenny around the house, Pete practically glued to her side.

"This is your room, I bet," Pete says, as we look into what any moron can tell is a girl's bedroom.

"No," she says. "That's Margaret's. I was hoping you'd meet her but . . ." I pick a magazine up off the window seat, drop it again without noticing what it is, and follow the two of them out of the room. "Andrew's not here either," she says, opening another door. All I can see is a huge bed. "Who the hell's Andrew?" I wonder. We go into the kitchen and Jenny makes us coffee. I don't say much but Pete chatters on and on. After coffee, that's it. We leave.

All the way back to Perth Pete is yakking away about what a cute chick Jenny is until, finally, I have to tell him to shut up.

"What's the matter with you?" he says. "You got a monopoly on every girl we meet?"

"Back off, OK? Just back off," I say.

"Listen, I wouldn't mind getting to know her a lot better, a whole lot better," says Pete. I feel like smashing him one.

"Well, I'm *going* to get to know her," I say. "So pick out someone else."

"Is that so?" Pete looks over at me, that dumb squirrel look of his. And then he starts to laugh.

"So back off!" I say. "Just back off!"

All night I'm brooding. "Did I say anything stupid?" I hardly said anything at all. She probably thought I was an idiot. "That's OK," I tell myself. "The next time you see her you'll . . ." It's barely light outside when I ring her number. "I'd like to come over and see you. Would that be all right?" I say the whole thing as if it were one word. "All right? All right?" Why doesn't she say something? She's going to hang up! Maybe it's the wrong number.

It's her voice, though, barely emerged from sleep. "Today, you mean?"

"Yesyesyesyes." I have to control myself so I don't go on and on. A pause.

"Sure," she says. "Come on over. We could have lunch." I feel my face beaming. "Hello?" she says. I say something. "But how will you get here?"

"I'll borrow a car. I'll take a bus. I'll get a cab. Don't you worry." I interrupt myself with a brief spell of idiot laughter. "I'll be right there." I hang up before she changes her mind and tells me not to come.

But how will I get there? The car of the night before has been returned and is not available today. A bus? That would take too long. A taxi. That's the only way. I don't even think of the cost.

Jenny's still in her robe when I arrive. "Come in," she says, and smiles. And I tip the cabbie and follow the swaying of her back until it settles itself in a chair and she turns around and I see that she's even more beautiful in the daylight. "I didn't expect you so early," she says. "Let me get dressed, then."

We fool around in the house till about eleven, and then I help her make lunch: a mushroom omelet, salad (lettuce from her garden, parsley, chives; the tomatoes won't be ripe for another few weeks), white wine, pears, coffee, and some little cakes she'd made and stuck in the freezer. She's a kindergarten teacher, I find out. Andrew's a good friend: her younger sister's ex-boyfriend. He needed a place to stay, that's all. She talks about the year she spent teaching on the air force base near Perth. Is this jealousy I feel? Her voice ripples with laughter. She tells me about her parents; about how happy she is to be living independently and not like her sisters; about how much she likes it here in this house where she can do what she wants and is only three minutes from her job.

Jenny's twenty-five. Her parents are Anglo-Burmese. She was brought up a strict Roman Catholic, lived in Burma until she was fourteen, when the family moved here to escape the

Communists. She loves people. That's the main thing in her life, she says: getting to know people, sharing. She's very suspicious of marriage. Trapped. She uses the word a lot. For too many people marriage is a trap. Early marriage, especially. She knows very few married people who are happy. When I tell her about my life with Jan, she nods sympathetically, seems to understand it all immediately. Love is free, she says, meaning also that it is beautiful, that it is mainly what human beings have to give to one another; but also that it is precious, not to be distributed lightly. It's the relationship that matters, though, not what you call it. Not what other people say about it, certainly. It should not have to be buttressed with words. It has nothing to do with laws. Her mother is the most loving person she knows. She lives for her husband, sacrifices herself all the time for her children. But she has never questioned anything. Jenny could never talk to her about ideas. Her father was very strict. She grew up afraid of him. " 'Till death do us part.' That sounds so grim," Jenny says. "It makes me shudder."

"Yes," I say, staring into her eyes. There are little fish of light that dart and jump there, that shimmer suddenly in flashing schools. I put my arms around her and press my lips to hers. She flows toward me; I sweep her up and we drift together—back and forth—this way, that.

"We don't have to get to know each other," I say. "We already do. Don't you feel that?" She laughs, but in a way that tells me she agrees.

In her room, on her bed, we play like seals—turn and glide and turn again; diving, rising in breathless leaps above the sea, plunging into it again. I look around me afterward. It's evening. Jenny is curled, asleep, beneath the wing of my arm. It is all so familiar, as if this were our room, our house. It's as if I have been away my whole life and have only just returned.

The next afternoon I'm back at Jenny's moving in my stuff. All Pete did was shake his head and smile when I told him. Already I'm thinking in terms of coming back here when the walk's over. It was different with the others, completely different. With Marie, for instance. She's the German nurse I met in France. We were crazy about each other. She flew to Ankara when we were stuck there for the winter and spent a month with me. We talked a lot about living together someday. At the same time, though, we never actually made plans. In fact, both of us knew, I think, that part of what made the relationship so special was just that it couldn't last. It was like loving someone you knew was going to die soon. Very intense. Every moment precious. But also sad. Built on sadness, you might say. Maybe you wouldn't even like the person if you could think ahead about it, plan a future. With Jenny, though, the future is already there. It's like the walk, in a way. The walk is there and I'm going to finish it, and Jenny's there and when the walk's over I'm coming back to her. No, it's more than that. The walk's still there but it's as if the whole reason for my going on the walk to begin with was so that Jenny and I could meet.

CHAPTER 13

"LET GO!" I YELL, AND PETE JUMPS CLEAR AND I'M RUNNING alongside, hanging onto the lead rope and just barely keeping on my feet. I can hear the wagon smashing along the fence. Pete's on the ground on the other side. I try to run fast enough to get my heels in and turn her, but as we pass the shed she picks up speed and I have to let go quick to get myself clear of the wheel. I hear the crash and don't even want to look up. It sounds as if the wagon's just been turned into kindling, and there's a terrible, saw-rasp braying coming from the mule. When I get my breath, I stagger over to inspect the remains. The mule's on her side, the wagon half on top of her, all four legs kicking. All the frame poles have been sheared off the wagon and one of the shafts is broken, but the rest, miraculously, seems all right. Pete comes limping over and we take a closer look.

"Damn mule," he says, but I just grunt and start trying to get her out of the harness. "You can't teach this mule anything," Pete says and glares down at her. "I give up."

Pete's been working with the mule for the past five days, but today's my first time with her. Probably it's my fault for taking her out of the corral, but I couldn't believe she'd be that skittish. Acted just like she thought the wagon was chasing her. "Get on up, dumb mule," I say, thinking that it's lucky she didn't break a leg.

"Bluey blew it again," yells Pete, as Bill Atwood comes walking over. Bill's a Mountie and our best buddy around here. He's the one who arranged to have the tourist bureau bring the mule to the police barracks, where he could help train her.

"Good news, mates," he sings out. "Guess who's in town? Monty Montana, Jr." We look blank. "From the States. He knows all about mules." Monty Montana, Jr., is a famous American cowboy and rodeo promoter. A week later, when we're ready to try again with the wagon, he comes out and shows us how to drop the mule when she bolts. Very simple. You tie a rope to each front foot, run it up through the collar and back to the wagon. If she starts to go, you pull her legs out from under her. By the end of the day the mule's pulling the wagon just fine. It's Monty Montana, Jr., who gives the mule her new name: Will Willie Make It II.

"Got a little horse in her," he says. "But she'll be OK."

"Good Willie," says Jenny, who adores our handsome blond mule. So we end up calling her that, too. Little Willie, big Willie, and now in-between Willie. For me, though, the real Willie is the one back at the Calcutta Polo Grounds. How I wish she were here! At least we found a good place for her. That much I'm glad about.

Things go faster now that we've got a mule and a wagon again, but I let Pete handle most of the details. All I can think about is that in a little over two weeks I'm going to have to leave Jenny. I wake up in the night sometimes and look at her curled there beside me and I wonder how she can sleep so peacefully, knowing we have so little more time together. I have to force myself not to wake her. So much can happen. There's so much farther to walk. The thought that I might never get back to her knifes through me and I feel my heart race and I am sure then that she will feel it too and wake up. But on she sleeps, hardly breathing; and for hours, sometimes, I lie there and think. I think about how lucky I am to have met her. "In this whole world to have met Jenny!" That thought alone makes me tremble. "What if they'd let us walk through Russia, or China? We'd never have met. What if Pete hadn't more or less made me go to that party? We could have been this close to each other and still I could have passed by without ever having seen her." I think of all the things that have led me here, to her; all the millions of circumstances that ended up bringing me to that party and not to some other place. If it hadn't been for John, for instance . . . Do I owe him this, then? Did he die so that Pete could come, so that I could find Jenny? No. I push that thought away from me. This has nothing to do with John. I've been through all that before.

Jenny laughs in her sleep. And talks. A lot. I never knew anybody before that did that. But she gets so serious, too, so mad when we start arguing about something. I want her to know everything about me. I want nothing to be hidden, so I've been telling her about the women I've met on the walk: about Michele, about that crazy girl in Ohio who put John and me up for the night. "You men, is that all you think about, sex?" Jenny says, her eyes blazing. She gets very upset, yet I must tell her. She wants me to, also. At least so she says. Yet every time I start, she ends by being angry. So why does she let me go on, then? It's the only subject so far we can't talk about, though, so I can't really complain. I wish I could tell her I didn't love those girls. But I did. And nothing that Jenny and I have between us will ever erase those times with them. But that doesn't mean I love her any less. It's just the opposite. If I'd met her straight from Minnesota, I probably wouldn't have known what to do. It was those others who taught me. But she won't let me go on. "You talk about it as if you'd been to a school," she tells me. "When women go out together, they don't discuss men that way." All I can do is drop the subject. I want her to know all of me, that's what it is: the good, the bad, the indifferent. I know her already. All she can reveal to me are the details. She's the crazy one, though she tells me I am. Crazy to have such naïve ideas about people. If she could walk with me for just a week, I'd show her.

"What about those dorks in Turkey shitting on top of the toilets in the gas stations?" I tell her. "You can't have any kind of relationship with them, whether you know the language or not."

"They're just poor and ignorant," she says.

"They're worse than animals," I say, but she shakes her head at me as if I knew nothing. I remember the first time—walking into that gas station men's room just outside of Istanbul. I couldn't believe it. A mountain of shit, piled three feet high at least, footprints where they'd climbed up to add to the pile. Bare feet! OK. In one way she was right. They wouldn't behave like that if they'd been educated differently. That was true enough. So what we should be doing rather than giving all those countries tanks and trucks and gas stations is educating them, making them all go to American schools from kindergarten on up, forcing them to drop all their backward ideas about religion and no birth control and . . . Wipe them out and start over again with the next batch, like that guy in Kabul said. That was probably the best idea. I couldn't say that to Jenny, though. Not unless I wanted a real fight on my hands. The thing is, though, I love her *for* her crazy ideas. I even wish sometimes I could think like that. But I can't. Even before the walk I was never that naïve. There are more bad people in the world than good ones, that's for sure. The walk proved that to me beyond any doubt, if we're going to talk numbers, that is.

I lie next to Jenny and think about the walk. What keeps coming back is why not give it up? Why not stay here with her, get a job, become a citizen? Or take her back to the States? Why go on? Just to get in the *Guinness Book of World Records*? Every step I take from now on will be a step away from her. She laughs when I say such things. I don't think she's really listening, even. It makes me angry, a little, that she seems not to worry about me.

    I lie there next to Jenny and realize that in no time now it will be Pete I'm curled up around. The stinking wagon. The endless miles. The sore feet. The baking heat. It's hotter on the Nullarbor Plain than any place on earth, we've been told. And we'll be crossing it in midsummer, one thousand miles of it. Desolation. Like the moon. And in the end, who is to know? It may be all for nothing. A tree can fall as you walk under it and kill you. . . .

    Jenny turns toward me in her sleep. I kiss her face: her lips, her eyes, her slim long throat; I press my nose into the fragrance of her hair. She murmurs and strokes me in return. Later, in the midst of love, even then she is not entirely awake. I hold her tightly in my arms and look beyond her to where the dawn light comes. Oh, if only I could take her with me. If only we could go together. If only we two were beneath one, tight skin.

CHAPTER

14

WILLIE'S PUSHING OUT SO FAST WE HAVE TO SET THE HEAVY-duty brake so we can keep up with her. It's eight o'clock the morning of January 3 and we're on our way to the Council House in Perth and the big send-off. The lord mayor, the media, the UNICEF people, they'll all be there. There are five mounted policemen prancing ahead of us: our escort into the city, thanks to Bill Atwood. Bill's walking with us. "There's nothing to it," he jokes, so I bet him a case of beer he can't last the three days. I think of Esber as I watch Bill striding along on the other side of Willie's head. Esber had been dying to walk with us, too, like so many others. He was so cocky that day he got off the bus in eastern Turkey, but three days later his knees were killing him so badly I almost wished he'd fallen down so we could have thrown him into the wagon. What a stubborn bastard, though. He'd be a half a mile behind us, at times, but he said he'd walk with us to the border of Iran, and, by Muhammad, he was going to do it. Ten days he was with us, and he made it, too. I don't know how. For the last four days I couldn't look at his face there was so much pain in it. It's funny. You can never tell who'll make a good walker and who won't. Esber had never had any trouble with his knees before, and by the end they were swollen so badly he had to cut slits in his pants. On the other hand, there was a girl named Irene who walked with us for a week, and she didn't seem to be bothered at all.

The lord mayor's the first to sign our new scroll. The others we sent home before we left Calcutta. There won't be too many signatures in this country, so it's good to have the first one be by such a big shot. There aren't more than a dozen places you could call towns the whole way across to Sydney, probably not half a dozen mayors. It isn't like Spain or Portugal or France where every little village has its string of officials. If you couldn't get the mayor, there was always his third cousin, the deputy so-and-so, who could sign for him. Here, what look like places on the map are mostly just truck stops, apparently. That's OK with me. It's pioneer country. Like the Old West. The bus company's even called that: the Ansett-Pioneer Bus Company. There's only one road, too—the Great Eastern Highway. That's good for us, though. The bus company's going to drop off stuff as we need it: water, food for Willie, even the mail. The heat'll be bad, but it can't be any worse than southern Afghanistan, I tell myself, and we're not going to practically die of dysentery from drinking putrid water in this country. And there'll be no dorks. Think of that! Just nice, friendly Aussies—one every fifty miles.

The TV camera crews follow us out of town and all afternoon people stop. "Luck to ya, mate." Ice-cold beer. Ice cream. It's like walking through the eastern United States all over again. Ken Colhung, the aboriginal cultural coordinator of Western Australia, pulls over and presents us with a boomerang, a gift from his people, he says. He is the first aborigine we have seen and we are touched by the gesture. It is as if Chief Crazy Horse had blessed us on our way. Why haven't we seen more aborigines? we ask Bill. Where are the kangaroos and the koala bears? Bill laughs. "Go up to Alice Springs. Plenty there." But that's in the middle of the Northern Territory, one thousand miles from here. We have seen one dead kangaroo beside the road. They don't run around all over the way you think. And they don't look all that different, either. If we hadn't seen one in the Perth Zoo, I might have mistaken it for some kind of small deer.

Toward evening we leave the lowlands and start to climb into the Darling Range, and Willie, who has been very strong all day, slows down to a crawl. We barely make it to the top of a long hill just this side of Glen Florry, and when we finally do, we decide to camp right there.

Jenny is due any time now. She's bringing out our dinner and will spend the night. For the next week or so—until it gets to be too far—we are her home and she will commute. Weekends she will spend with us. We are not going to say good-bye, we have decided. Not till I leave Australia, anyway.

Bill, who sits with his feet in a basin of hot water after we eat, says that Willie needs shoes. We had hoped that her hooves were tough enough so that wouldn't be necessary, but the road has worn them down considerably already. So we'll stop in Mundaring the next day and see to that. A big disappointment. There won't be many blacksmiths on the Nullarbor Plain, that's for sure.

What with the blacksmith stop and Bill's feet, we make only fifteen miles the next day. To a place called Lakes. A Shell truck stop; that's all it is. But the Aussie that runs the place is a "sporty bloke," as they say down under, and not only invites us in for a free meal but loads us down with beer as well. The next morning I win my bet with Bill. He calls his wife to come pick him up early. Riding around the world on horseback; that's more his style, he says. That night we notice that Willie has already developed harness sores. Partly to rest her, but mostly because it's Sunday and Jenny's with us, we stay where we are for the day, just goofing around.

The whole next week we take it easy. It's good walking weather but neither Pete nor the mule is up to it. I don't mind too much, but by the time Jenny gets back to us late the following Friday night, we're in the wheat belt and everything's changed. It's baking hot. Every day the temperature's over one hundred degrees. Friday, there's a furnace blast in our face all day, and we make only seventeen miles, even though the country's completely flat. The worst thing, though, is the flies. No place on the walk have they been as bad as this. They don't sting. They don't bite. They simply occupy every square inch of exposed flesh, searching for water. If we forget and open our mouths to breathe, they try to crawl down our throats. Our nostrils, the corners of our eyes, anyplace they find a little moisture, there they are, inches thick. Swatting them continuously we know now what is meant by the Australian salute. Fortunately, we have gotten word back to Jenny and she brings us fly nets that fit over our hats. One for Willie, too. The nets make the flies bearable, but now we walk as if blindfolded. Flies cling to the mesh, burrowing to get in. Our faces pour with sweat. And the flies get inside, some of them, so there is the continual slapping, squeezing, squashing of flies up against wet skin. Even the wind fails to drive them away. It's torture to eat our meals, to drink a glass of water. The only real relief is at the truck stops. Inside, drinking a beer, we are free of them for the moment. The Aussies here laugh at us, though. They wear no nets, claim not to be bothered by the flies. "They're worse than the dorks," says Pete. Almost, I agree.

The flies disappear at night, thank God, but during the middle of the day, when we stop to rest in the worst of the heat, they descend in even greater numbers. We sleep with our nets on. Flies crawl inside our clothing. Pete or I will suddenly get up, unable to stand it any longer, rip off a shirt, a pair of pants. "There. Got you, you bastard." But ten more get in, twenty more. We never get used to them, but rising to our knees in the wagon, screaming with fury, we find twenty minutes have gone by since we were last awake. We learn to live with crawling things, like people infested with lice, or crabs, or worms. Still, I will be lying there, dreaming of Jenny, and suddenly snort to attention as a fly crawls into my nostril. Pete wakes up and it takes maybe ten minutes to settle us down again. In a way the flies *are* worse than the dorks. They are with us from sunup to sundown, and there is no satisfaction in killing them.

"Come on," I say to Pete one morning. "Let's go!" He just looks at me as if the sun's boiled my brains at last, and we continue on at the same maddeningly slow pace. Will Willie is not like Willie the Second. She can't go fast, except in spurts, and then we pay for it because she goes much slower again afterward. No stamina. No guts. No good old Portuguese mule strength. All day we trudge along through wheat fields that are as vast and empty as the sea. Swatting flies off the net is habitual now and I am separated a little from myself, riding a foot or so up in the air and thinking about how wonderful my life will be when all this is over.

On January 17 we leave the wheat fields and enter the bush country. At first it's a relief. The ground rolls a little and we can see to the horizon line in every direction. Gradually, the bushes become small trees, though. Packed close together they form a wall ten to fifteen feet high on either side. It's like walking through the villages in Turkey or Iran. The road is a corridor, an oven, a dry riverbed left over from some ancient, doomed landscape. Its perfect straightness makes it seem all the worse, as if all the days of one's life could be set out in a line, measured off, considered. From the twenty-foot top of a hill I can see into next week. From Southern Cross to Kalgoorlie is 118 miles: not a house, not a person anywhere. Only a water pipe; a silver snake three feet in diameter built to keep the gold miners of Kalgoorlie alive. A smaller pipe continues 106 miles farther to Norseman, on the edge of the Nullarbor Plain.

If it's this bad here, what will it be like there, where there is nothing? Not a tree. Not a bush. Seven hundred miles of moonscape nothingness, where the sun bakes down and frizzles whatever crawls out onto it, like so much bacon. For 250 miles there is no paved road even, nothing but a dirt track across a salt lake. Even here the dust swirls and is gritty between our teeth and tastes of gunmetal. Our lips are cracked and sandpaper dry. Every half hour or so trucks pass us three or four together: a caravan of trucks and cars in a sudden stream of dust-choking traffic. And then the silence closes down again, and there is only the creak of the wagon, the clomp of Willie's shoes against the rough, metallic surface of the road, and the squeak of harness leather.

For the next three days we plod along, looking straight in front of us, as if the wall of bushes on either side were blinders. I have grown used to the mat of flies on the wet back of my shirt. Willie they cover entirely: harness, shaft, net, and all. Black Willie, no longer blond. They are more like bees than flies. They burrow. They crawl. Not one flies off when Willie twitches her head up and brays. I make up letters to Jenny as I walk along. I talk to her. I tell her everything I've ever done, starting as far back as I can remember. It's funny, the farther away I get from her, the nearer she is to me, in a way. The boredom of the walk helps. I curse at Pete when he yells at me or stops us for a rest, when he takes me out of it. He's taking me away from Jenny, my Jenny. For an hour, maybe, I've been with her, really been with her the way you are in dreams, and he's forced me back into this hellhole, this fly pit. I hate him for doing that and tell him so. We talk very little now, even after we've stopped. I have nothing I want to talk to him about. The best times are the long, late afternoons when we trudge along in silence and it's a little cooler. Pete and I have already had a fight, probably, so he's not looking for conversation, and I'm off in my own world with Jenny. It's better even than at night. Night dreams you can't control.

Toward the end of the week—it's been almost three weeks since I've seen Jenny—we get near Kalgoorlie and the countryside opens up. The land rises, then falls away again: a copper-colored, sandy sea. Off in the distance there are hills, the gold mine hills of Kalgoorlie. All afternoon they rise before us higher and higher, and in the evening we are there. "God, how I want a shower," says Pete. It's been six days. We're planning to spend a week here, get a good rest. Pete's feet are killing him. Willie is barely able to make it into town. As for me, though, the thought of spending a week here in this ghost town, just sitting around, is enough to drive me crazy. That night, in the dining room of the Australia House, we're sitting around talking to a blacksmith we've met about crossing the Nullarbor, and he tells us that what we need for Willie are shoes made out of barium. "Last one thousand miles," he says. "The only place to get 'em, though, is in Perth."

"I'll go," I say. "I'll leave right away. Tonight. I'll hitchhike."

"Good," says Pete. "You do that. And stay as long as you like." He says it kind of nasty and the other guy looks over at me and gives a funny laugh, but I hardly notice either of them, I'm so excited. I leave right after dinner and almost immediately catch a ride all the way to Perth. At two in the morning I'm pounding on Jenny's door.

"It's me. Dave." She laughs in sleepy surprise and lets me in. Now that I'm with her again it is as if I never left. The walk is a nightmare I wake up from in Jenny's arms. She has to teach during the day, but in the early morning, in the evening, and all night long, we are together. I go shopping for us. I get Willie's shoes made up. One day I go to the beach; but most of the time I spend in and around the house. This week it is our house; and transplanting bushes, waxing floors, making the bed, even, all of these things fill me with pleasure. The central room is twenty feet by forty-five feet, is made of stone, has a two-story-high ceiling with exposed beams, and at the back is a huge fireplace. It was part of a monastery, originally. From the porch we can see

Perth and the glistening ocean beyond. It is airy and spacious, this house, yet cozy and warm. When Jenny comes into the driveway, horn tooting, scattering the pebbles, I run to clasp her in my arms. Finally, I must leave her, though; make the long trip back to Kalgoorlie.

Early Monday morning, on her way to school, she drives me to the Great Eastern Highway.

"April," I tell her. "Early April." If we haven't gotten to Sydney by then, God help us. Jenny has two weeks off in May. We try to say good-bye and cannot even look away from each other. Finally, gently, she makes me leave. She is already late for her class. I walk away backward; dead, numb inside. She waves to me from the window of her car. The traffic separates us for a minute. I want to hold that picture of her smiling face in my mind just as it is until I return. A car stops and I get in. I do not get a chance to wave again. When the traffic passes she will look across and not see me. For a second she will wonder where I am. So it's that momentary frown before she realizes what has happened that I take with me.

"Going far?" the man says.

"Around the world," I answer, absently, and he laughs. The deadness within me has returned. Sydney is a continent away. And the end of the walk? On the other side of the earth. I stare out the window. Will I ever see Jenny again?

CHAPTER 15

"No I won't," I say, and walk off and leave him there. Willie is moving out at a good steady clip, for once. What does he want me to do? Hold her back? Put on the brake? Just because of his damn feet? If his feet are still giving him trouble after all this time, he'd better quit pampering them and just grin and bear it. It's a good hundred yards before I turn my head to see what he's doing. He's still sitting there beside the road, massaging his dumb foot. "The hell with it," I say out loud. "I'm not going back and I'm not going to wait for him, either. Let him catch up." Half an hour later I stop and look back down the road but I can't make him out. "Probably he hitched back to Norseman," I say to myself, and Willie and I move out again at the same fast pace. Two more miles and I stop again. It's almost dinner time, anyway. I'll stop here, get everything ready; and then, when he finally walks up, he'll just have to eat and run, that's all. We'll walk late to make up for it.

I've fed and watered Willie, finished my half of the stew, and I'm drinking a cup of coffee and trying to ignore the flies when a car drives up and out gets Pete.

"What's all this?" I say. But Pete doesn't even look at me. He's saying something to the people in the car.

"What are you doing?" I shout. "If you think I'm going

to go back and walk that part over again, just for you, you've got another . . ."

"You don't have to," he says. "I've quit."

"What do you mean, you've quit?" I say, and walk over toward him.

"I've quit the walk," he says, and limps over to the back of the wagon and sits down. "That food for me?"

"You can't do that," I say. "You can't just up and quit. We haven't even talked about it. Where do you get off, just quitting like that?"

"I've been talking about it plenty," says Pete. "Are you going to hand me that stew or do I have to hobble over and get it?"

Suddenly, I don't know what to say. How can he quit the walk? Here, just when we're about to go into the Nullarbor, he's going to take off and leave me?

"All right," I say. "Eat this." I start to hitch up Willie. "I hope you marked the spot."

"There's no point going back," says Pete. "The walk's over."

I look at him. Can he be serious?

"I wanted to do it for John," he says. "I wanted to say that two Kunst brothers finished the walk, but it's too late now."

"You want to rest a day or two? Is that it? We'll go back to Norseman, take it a little slower."

"No," says Pete. "I'm going to ride in the wagon from now on."

Actually, it works out pretty well. Pete sits in the wagon and flicks Willie on the ass when she needs it, and I'm able to go at my own pace. Pete's bored as hell, though. The second day out from Norseman he starts writing this dumb poem: "The Ballad of the Kunst Brothers' Walk: One Step at a Time," he calls it. "Listen to this," he says, and I have to yank my mind back from thinking about Jenny, or concentrating on that line of road in the distance, or on leading Willie.

> "Two Minnesota brothers put their feet down
> As they walked down out of town.
> John went merely for the search,
> And Dave left behind the social skirts.
> The mule is named from a nun
> One step at a time, Willie, don't run."

"Great," I say. "Terrific." It goes on and on. Day after day. "What about this? Does this sound better?" Pete's no poet, that's for sure.

We're making thirty-four, thirty-six miles a day, and then one morning I put my right foot down on something the wrong way and I feel my ankle give. By noon it's all swollen up, and it's all I can do not to yell out each time I put my weight on it. For the first time on the walk I'm glad when the mule starts to slow down. I can't tell Pete what's happened, of course. I won't even let myself limp. He keeps beating Willie on the ass and I tug away at her, too, and pretend I'm mad; but good old Willie just settles back to a crawl and stays there, God bless her, no matter what we do. Sunup to sundown. Sixteen hours. Twenty-two, twenty-five miles a day. All the people on the walk who had problems with their feet. I'd hear every one of them laughing till the day I died if I even let on I was in trouble. When that truck tire spun a rock into John's ankle in eastern Turkey, I wouldn't let him stop. Hell, he was begging me, and I wouldn't; not even after I saw how swollen it was. "Walk it out," I kept saying. He kept walking, too, because he knew I'd go on without him if he didn't. Most of the time he was either crying or cursing at me, but three days later the swelling was down and he was walking normally again. I was always having to do stuff like that with John. Pete, too. Esber was the only one who was so hard on himself that I didn't have to push him. With John and Pete it was always, "At the top of the hill. We'll stop there. Count to one hundred. Count to 100 five times. Count it again." That's the only way I'd keep them moving. So now I do that for myself.

In a funny way the flies help. They're worse than ever since we've gotten into the Nullarbor, but what I do is play this game. A fly gets down inside the collar of my shirt and I feel him crawling toward my right armpit, so I say, "Go ahead. Enjoy yourself, little fly. Suck away." And I lift my elbow to make things easier for him. When I feel the fly settle under my arm, I settle myself down, too. I imagine him in there, sucking away at my sweat and having the time of his life. What a paradise, right? This little fly, out of all the other flies there are in the Nullarbor, only this one has made it to the big oasis. How happy he must be! In ecstasy! I see myself for a moment drinking at my own armpit and that makes me laugh, and all this time I haven't thought once about my ankle and how much it hurts. The trouble is I can't always do this. A fly will get in and everything'll be fine until he's almost there. Then something happens. I step down harder than usual; Pete yells out part of his poem, maybe; Willie picks up speed for a little while. And then, instead of raising my elbow politely for my visitor, I grind down with it and try to crush the little bastard up against my rib cage; I reach over with my hand and try to squeeze the life out of the little sweat sucker. Part of the time I'm walking in a daze, and then I'll suddenly wake up to the pain. I couldn't tell you where I've been or for how long. And sometimes there're voices talking to me: John or Pete or someone else who's been on the walk with us. I'll hear Pete yell out, and realize I've been talking out loud. Sometimes, I think I must black out for a second from the pain. For four days it's like this. And then the swelling starts to go down. Pretty soon after that I'm all right again.

The Nullarbor's not as bad as we expected. It's not as hot, for one thing. The nights are actually chilly. And there're small trees: some shade. Even grass in places. And then people have been stopping and giving us things: fried chicken from our friends in Norseman, cold beer, ice cream once from the driver of a refrigerator truck. And every other day there's the bus. The people all get out and talk to us and take pictures, and that breaks up the monotony. And then there are Jenny's letters. They fill the days for me. It doesn't much matter to me where I am. It's as if she's with me; or rather, I'm with her. I ride with her into Perth. We go for a swim. We lie on the beach drinking Swan beer. There's a little girl named Sally in her class who sends me a picture of us walking with the mule. Pete pins it up in the wagon. Jenny's whole class is following the walk every step of the way. She sends me newspaper clippings about how we're doing. Now that Pete's quit the walk there is much speculation about whether or not I will do the same. "Not on your life," I tell her, though I hardly have to. She sends us homemade cookies. On Valentine's Day the bus drops off a package that contains not only all sorts of goodies but a handmade valentine from each of the children in her class as well as an illustrated scroll of our walk from Perth: the class project for the past month.

That night Pete tells me that when we get to Ceduna, on the other side of Nullarbor, he's going to fly home. Well, I figured something like that. He'd promised Nancy he'd be back within a year and by then the year will be up. OK, I tell him. Ceduna is only halfway to Sydney but from there on it should be a piece of cake. I don't even try to argue with him. He's quit the walk, hasn't he? He's given me a year. Gotten me through Pakistan and India, and if he stays for the rest of the Nullarbor, he's done more than his share. I don't even feel badly about his quitting now, and I tell him so.

The Nullarbor's quite beautiful, in its own desolate way. Pete doesn't agree, but for me the very fact that I know that there's nothing around me for hundreds of miles makes it special. It's like areas of Afghanistan that we walked in, except that it's ten times as big and there's no one around: no bandits, no killers to worry about; no ignorant, stupid dorks, either. It's flat and open, just scratchy patches of small bushes and a coarse kind of grass. In Afghanistan or Turkey, say, there'd be mountains way off on the horizon. Here there's nothing. It's as if you were looking at the ends of the world. At night, the stars are so bright you can see almost as well as in the daytime. You look up and there, right in the middle of the sky, is the Southern Cross. Most nights there's not a sound, either; no wind. It's as if you're the only person in the world and you're standing right in the middle of it, and up above you are all the other worlds, maybe some of them with life on them, too. You stand there looking up, and the stars are so bright your head seems to be right up in them. Your head's touching the sky. You stand alone and tall and there's no one else in the whole world that you can see.

On February 21 we come to the end of the paved road: 253 miles of dirt. Right away Willie slows down to a crawl. It's like walking on the shoulders all the time—the wagon pulls twice as hard. Pete has to get out and push every so often, but even with that help Willie doesn't do much. She's got a bad harness sore, for one thing, but she must have worms too, because all her pep seems to be gone. She's not eating her feed and she's getting thin and scrawny-looking and her head's down. We ask the Pioneer bus driver to drop us off some worm medicine; but even if that works, neither Pete nor I think Willie's going to make it across the rest of the Nullarbor. It's her eyes. Old Willie the Second's eyes used to get red-rimmed and wild-looking. The worse things got, the fiercer they got, even when her head was practically on the ground. But this Willie's eyes just look sort of washed out. They're colorless, almost. They never were much, but now they look as if there's nothing behind them at all: no spark, no anger, even; nothing. "Get on up, you mule," says Pete, and cuts Willie with his switch. Only the flies move, and before Pete can raise the switch they are back down again. The road is not just dirt, either. It's made up mostly of small stones, some of them the size of billiard balls. Willie picks these up in her hooves, which causes her to go lame if we don't get them right out. I don't dare take my eyes off the ground because I might step on something and twist my ankle again, and that means I can't think about anything else.

We've been walking for two days on this road when the weather turns blistering hot again and a scorching wind comes up from the east and the dust starts to blow. Pete has to push most of the time now just to keep us moving, and it's like it was in Afghanistan when the winds blew there in the southern desert: sand stinging our faces, can't see past Willie's nose most of the time, can't look up at all, sometimes; have to stop again so as not to lose the road entirely. Only here the sand is finer: a light brown dirt that covers us in no time. We can scarcely breathe, must turn our backs; blind in the choking dust storm. We huddle behind the wagon, waiting for the wind to stop howling, sometimes for hours on end; or we plod on again, into the choking, stinging dirt. People stop and help us, though. Most of them have heard about us and they bring us things to make our trip easier. Once in a while a car stops that is air-conditioned, and we get in, have something cold to drink, and are better able to take the punishment when we get out again. Willie, though—there isn't much they can do for Willie. When the dust is bad like this, we don't make more than fifteen, twenty miles a day. The dust storms come and go. We'll see them on the horizon, heading for us, or they sweep past the other way. Someone brings us masks to breathe through. He even tries to fix up something for Willie, but can't get it to work. The nights are cold, now.

All along this stretch there are discarded tires and we use them to build huge bonfires. Generally, the winds go down in the evening and after we clean up a little and eat, we'll sit by the fire and talk or just look out into the night. One night, in particular, stands out. A full moon, so large it half fills the sky. It's setting and, as we watch, it goes from white to yellow to orange and finally to a deep, blood red. For a moment it sits there on the edge, then flattens out and slowly sinks—dripping, punctured—beneath the earth; and the sky is lit up again by the sharp, white arc lights of the stars.

Thanks to an ex-major who has spent much of his army career in India with mules, Willie's sores are somewhat improved. "Piss on them," he says. "Two, three times a day." So we drink up and wash down Willie with our urine, and it does seem to help. We have a new problem, though. Willie's hooves are starting to crack from the dryness. I rub on Crisco, peanut butter. Nothing seems to do much good. The worm medicine has gotten her to eat better again, but we're still not covering more than twenty-five miles a day. Yet each day is a day nearer Ceduna: the end of the dirt and the worst of the walking. When he's not pushing, Pete works on his poem:

> The Afghan desert it just wouldn't cease,
> Americans sent warnings, "Your life is on lease."
> In the Kabul Gorge, where bandits did lurch,
> John, shot through the heart, lost in his search.

On March 4, around sunset, we reach the hard-top. What a difference. Even Willie perks up for a while. It's only two days to Ceduna, now. That night we polish off the bottle of booze we've been saving to celebrate the end of the dirt. The plan now is that we'll leave Willie in Ceduna to rest up for a couple of weeks and I'll go back to Perth while Pete goes on to Sydney by bus and then flies to Los Angeles. So this is sort of good-bye, too. We're both too excited for it to mean very much, actually. Pete'll be gone. OK. I'll miss him, that's for sure. But all I can think of now is that in a few days I'll be with Jenny again. And all Pete can think of, I guess, is that pretty soon he'll be back home and surrounded by all the comforts.

CHAPTER 16

IT'S BEEN A LONG TIME SINCE I WALKED ALONE. THE WALKING part's all right, but at night, after the mule's settled and I've had my supper, I feel empty in a way I never have before. After almost a month with Jenny, I'm spoiled, of course. But it's not that. It has something to do with the country, I think. I'd thought that after I left Ceduna there would be more people. I'd be able to stop someplace for supper, maybe get put up in a motel, have a few beers, talk to the local Aussies some; but three days out of Ceduna I'm still in the middle of nowhere. I miss Pete. I miss him like hell. I miss his dumb poem. I miss him in the wagon at night: his laugh, his bony body, his smell, even. Missing Jenny is partly a pleasure. I'm with her in my mind. We're not really separated. With Pete, though, it's strictly pain. I need him, damnit. He should be here. I'm all alone with nothing but a dumb mule and sometimes I want to howl, howl like a wolf; or cry, or hit something.

For the first few days Willie moves right out. She's in good shape, better shape than when we first got her, I think. I'm doing my four miles an hour and we're making thirty-three, thirty-seven miles a day. Then she slows down. Just before noon on the third day, she slows down to about half-speed, and from then on that's it. I try hitting her on the ass with a switch the way I used to with Willie the Second, but Will Willie won't move nohow. When we stop that night, later than usual, we've made only twenty-four miles. There's something wrong with her but I can't figure out what it is. She'll go along all right for a while and then, for no apparent reason, she'll suddenly slow down to a crawl. It's infuriating, like driving a car with bad plugs. Or she'll stretch herself out like a pair of scissors and spend five minutes farting. Maybe there's something wrong with her stomach, I don't know.

It's April 14, eight days since I left Ceduna, and I'm camped by the side of the road fifteen miles from Port Augusta, the second-largest town in South Australia. I'd been hoping to get into town that night, but Willie just can't make it. I take the harness off her, and I walk her around some while she stretches and strains and rattles off her farts. Then I tie her up to the wagon for the night. Tomorrow's Easter. Some Easter. In town there'd be things to do. I could celebrate a little, but out here it isn't much different from being on the Nullarbor. I treat myself to a little brandy that night and eat two of the Easter eggs that Jenny has sent out to me via the Pioneer bus lines. One's painted in wild circles of colors to look like the sun, I guess, and I hate to crack it. The other's covered with tiny chicks and flowers and bunnies. I eat them both and feel better. Tomorrow, at least, this part of the walk'll be over. From here on it's *got* to be more populated. I write Jenny a long letter and then decide to write Pete, too. I end up writing to the kids and my folks as well. What an evening of letter-writing, something I seldom do. I go to bed early and, for once, fall right to sleep.

Just after midnight I'm wakened by Willie's jerking at the wagon with her rope. Goddamn mule! What is it now? I get up and go out and untie her and start to walk her around. She's farting and stretching, and even in the darkness I can see that her stomach's all swollen out. For half an hour I walk around with her. No change. Another half hour. There's probably something else I could do, but what? What I mainly think of is what a bore this is. "Walk all day and then when I want to sleep I've got to walk some more just to keep the damn mule from blowing up from her own gas." Another half hour. Willie's some better, it seems. She's stopped braying, anyway, and her stomach, when I smooth my hand over it, seems less swollen. I walk with her for another half hour and then tie her to a post by the side of the road. She doesn't want me to stop, of course. She brays and farts at me when I leave, but what can I do? I can't walk her all night.

It's six-thirty before I'm out of the wagon again. At first I don't see her. She's broken loose, I think, and curse myself for not tying her to the wagon. Then I see her. She's lying down, right near where I tied her. "Lazy mule," I think, and walk over to her. Still asleep! I kick her in the side. "Come on, Willie," I say. "Up!" But she doesn't move. So I grab her lead rope and give it a jerk. Nothing. No response. So much dead weight. And then I realize what's happened and my whole body goes cold. Willie is dead. Her stomach: swollen up to twice its size, her legs stuck straight out in front of her. I get down on my knees. Her eyes are glazed. She's not breathing. What I feel mostly is numb. I can't take it in. I walk around her body once and then just stand there looking at her and wondering what to do. I stare at her, then stare off into the distance, then look back and stare once more. Her tongue's hanging out of her mouth and I notice that it's thick with flies. Flies swarm over her eyes, in and out of her nostrils. Her round, swollen belly is black with flies. I turn her over on her other side, I don't know why, and there is the long, sad sound of escaping gas. I don't know what to do, what to think, even. I stand there watching the flies. Finally, a car drives up and a man gets out and comes walking over to me.

"Tom Truslow," he says, and takes my hand. He's a big,

Senator Hubert Humphrey wishing us well.

The St. Paul Dispatch and Pioneer Press

Pathan tribesmen in the Khyber Pass.

With the Khyber Rifles.

Chitral, Pakistan: Pete and Dave with Mt. Tirich Mir in the background.

National Assembly Building, Islamabad, Pakistan.

Crowds in eastern India.

Trudging through the Indian monsoons.

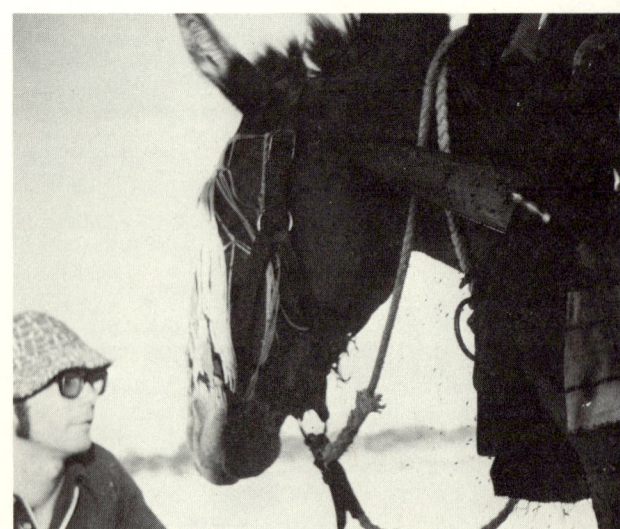

The flies of Western Australia.

Sunset: The end of the dirt road on the Nullarbor Plain, South Australia.

Jenny and Dave.

Jenny at the wheel of Will She Make It.

Newport Beach, California: With Jerry Patterson, the Mayor of Santa Ana, and my new mule, Willie Will Make It.

WIDE WORLD PHOTOS

Dave checking out his shoes before leaving Santa Ana, California.

WIDE WORLD PHOTOS

The home stretch: Iowa.

Waseca: The end of the walk. From left to right, Mr. Kunst, Pete, Danny, Willie Make It, Rich Ebensteiner, Dave, and Bradley.

WIDE WORLD PHOTOS

"I wave my hat and the crowd cheers and shouts hurray."

The Minneapolis Tribune

burly farmer type with a red face and a thick shock of steel-gray hair. "Poor fella," he says, looking down at Willie. And then he walks me over to the wagon. "We can tow you over to my place," he says. I look up at him. What for? He backs his truck over and ropes the shafts to the sides. "Get in," he says. "It's just down the road."

"What about her?" I say, as we drive off.

"Don't you worry. The road crew'll take care of her." I nod my head. Why is it that what I feel mostly is nothing? No sorrow. No tears. After Willie the Second I guess I just thought mules were unkillable. But Will Willie wasn't all mule, was she? That's it. She was part horse. And used to carrying a pack, not pulling a wagon. You had to be a mule to take this kind of thing. You had to be tough and stubborn and ornery and mean sometimes, or you wouldn't make it. And where am I going to get a good mule, now?

The Truslows give me breakfast and then I hitch a ride into Port Augusta, where the police put me in touch with Dave Tyler and Brad McNally of radio station 5 AU. They're great guys. Real good-hearted Aussies; and that very afternoon we drive out in their mobile unit and tow the wagon to the Flanders Hotel where I've been given a room. What with the radio station and the local paper, I'm getting quite a bit of publicity as I walk into town at four miles an hour next to Dave and Brad who are towing the wagon. For the first time since leaving Perth there are crowds around me, people waving. "At a go, Yank," someone yells. A dozen voices pick it up. It feels good. And then I get my big idea. Boy, am I excited. As soon as I get to my room, I call Jenny. I break the news to her about Willie; and then, when she's over the shock of that, I tell her what I've just thought of. How would she like to drive here and tow the wagon for me the rest of the way to Sydney? I'd walk next to her and we could talk. We'd be together at night. It would be boring for her but it was less than a thousand miles. We could probably do that in three weeks. Could she get time off—that much time—from school? That was the big question.

Yes, but not till May 11. Jenny is as excited as I am. What a lark it will be!

"My good mule," I say.

"Can you come home right away?" she says.

"The next bus," I tell her.

We're both laughing now. Everything seems so simple all of a sudden. How come we never thought of this before?

CHAPTER

17

"It was right about here," I say, and then I see them and I try to distract Jenny, but it's too late and she lets out a cry and buries her face in my chest. She's shaking so that it's all I can do to keep the car under control till I get us off the road. She's exhausted, poor kid, but all I can think of is that idiot road crew. Letting the hooves stick out like that. Couldn't they have buried Willie a little deeper? I'm pretty shaken up myself, actually, but mostly from all this driving. Here it is Tuesday morning, May 14, and we've been driving steadily since Sunday afternoon. Those floods in the Nullarbor held us back half a day at least. Lucky it wasn't like that when Pete and I went across.

When we get to the radio station, Jenny finds a place to go to sleep, but I'm too hopped up, so Dave Tyler and I get the wagon hitched up to Jenny's car and we take a practice spin.

"How far is that?" I say, after half an hour.

"Just two miles," beams Dave Tyler.

"It's my natural pace," I tell him. "No problem." We've got twenty-four days to get to Sydney before Jenny's vacation time is up; so we'll have to average a little over forty miles a day. I'd like to leave that afternoon, but Jenny's still dead to the world and, anyway, the roads out of Port Augusta are badly flooded. I'm dying to get going, though. It's going to be so good not to have a damn mule to mess around with anymore. So good just to walk out at my own natural pace for once, not to have to think about anyone else—about Pete's feet or John's feet or anyone else's feet except my own feet, which are fine. And to top it off, Jenny'll be with me, in the car.

All the way from Perth it was like riding one long wave. I was too excited to be tired. There Jenny was sleeping next to me, her head warming my lap. Nothing but the two of us purring across that open, flat country; the stars so bright that half the time I'd drive along with the lights off. It was more like traveling through space than along a road. The hum of tires, the long, straight road, the steady drone of the engine. I swear I felt rested in the morning. Yesterday we talked a lot about what I'd do when I finished the walk, how I'd come back here and write a book.

I'd tell everybody I was going to California to hole up in the mountains so I could get it all down and wouldn't be disturbed, but what I'd really do is come straight back to Perth and live with Jenny, and every day I'd write and she would teach and on her vacations and on weekends we'd travel or go to the beach or just stay around the house by ourselves.

"It'll take me at least a year," I said. She looked over at me and smiled.

"And after that?"

"Who knows? Does it matter?"

"No. Not really."

"Hey. I love you."

She laughed. "Why do you always say that? I know."

" 'Cause I do. And I want you to think about that and nothing else. OK?"

It would take a year to get the divorce. But I couldn't let on I was coming back here. One thing a woman can't stand is knowing there's another woman on the scene. Pete would have to cover for me, that's all.

"I just can't believe you walked all this," Jenny kept saying.

"I didn't," I said. "I ran most of the way."

"Sure," she said. "And that's Los Angeles up there ahead, isn't it?"

"You mean that shiny place?" I said, and I spun a beer bottle off into the Nullarbor with a flick of a tire.

"Are you a good camper, too?" says Dave Tyler to Jenny, over dinner at a restaurant he's taken us to.

"I will be, I think."

"Sure she is," I say. "Anyway, all she has to do is drive."

It's not till the following afternoon that we finally leave. Even then the water comes almost to the top of the wheels in places. Once we get past the Murray River bottoms, though, the land dries out and, except for a strong head wind, the walking's good. I walk along without stopping all afternoon and by dark we've made exactly twenty miles.

"OK," I say, leaning in the window. "We can stop here." Jenny nods and pulls the wagon off the road. She gets out and stretches her arms up over her head, and then brings them down in front of her and touches the ground. Up and down, a dozen or so times.

"Tired?" I say, and walk over to her.

"Stiff, mostly," she says. She looks tired, though. There are circles under her eyes. "These flies," she says, and swats her face. Inside the car, with the windows shut, she hasn't been bothered.

"You get used to them," I say. "No choice."

"Get me some water, Dave?"

"You'll be all right," I say. "I'll get dinner ready."

She doesn't eat much of the stew, though, and as soon as we finish we set up the bed. It's only about nine but she's asleep by the time I've got the light out. A car goes by and I feel her tense up in my arms.

"It's OK," I murmur, but she doesn't wake up.

I want to tell her how much I love her but she sleeps as if drugged. I content myself by hugging her close.

Jenny is better in the morning. We get off about eight and for a while we talk as I walk next to her. The flies get too bad, though, and after she shuts the window I go off into my mind, the way I've been used to doing ever since the walk began.

"Let's see, feet. Can you miss that crack? Good. Stretch a bit. Did it! Shorten up, now. OK. That's an elephant. No, South America. Be careful, there're six in a row here. 'Step on a crack, break your mother's back.' Where did I learn that? School? Before that. Good old Ma. Taking me out to the woods with my gun and pack of food and shells—right there, waiting for me, three days later. Wanting to come on the walk like that. Meant it, too. Would have, if Dad and John hadn't laughed her out of it.

"How far to that pole? 140 steps? 147. Not bad. 31 steps equal 100 feet. 310 is 1,000 feet. 1,550 equal 5,000 feet. 1,643 is 5,280 feet or one mile. Seven miles then. To the top of that little rise? About half a mile."

"What?" I say. Jenny has the window open.

"Can we stop for a while?"

"Sure," I say. It's ten past eleven. We split a chocolate bar and drink some orange juice. We have a cooler, now, with ice in it. What luxury! It's warm in the car, but not unpleasant.

"My foot gets so stiff," Jenny says. "I don't see how you do it, just going on and on like this? Don't you get bored?"

"You get used to it," I say. "You OK now?"

Jenny leans over and I kiss her.

"Good-on-yer," I say, and we both laugh. Aussie slang!

We drink the rest of the juice and get out of the car. Jenny jogs off up the road a ways. "Dumbbell," I say to myself as I watch her. "Stop and let her stretch every hour. What's the matter with you?"

It's three o'clock in the afternoon. Our second stop since

lunch. Jenny is looking strained and tired. "What if I drive ahead a couple of miles and wait for you?"

"OK," I say. "Good idea."

John and I used to do this with the Afghan soldiers that were sent out to protect us. They couldn't keep up so they'd catch rides and then we'd meet up again. It was hopscotch like that all day long. Jenny drives off slowly and pretty soon she's out of sight. Funny to see the wagon just move off ahead of me like that.

Whoosh! The damn car practically hits me. A horn blares. I turn and get off the road just in time. I'm walking on the shoulder, now. Goddamn Aussie drivers. "Want a ride, mate?"

"No thanks," I say, and the man drives on again. Must think he met a real crazy.

"It's no good," I say to Jenny, when I catch her half an hour later. I could carry a sign on my back, I guess. But I still couldn't walk on the pavement.

"That's OK," Jenny says, and she goes back to driving alongside of me.

When we stop that night we've made thirty-six miles. Jenny hardly speaks all evening and goes to bed right after we eat. We'll have to do something, I realize. But what?

The next morning I adjust the idle so all Jenny has to do is put her foot on the brake every once in a while. Why didn't I think of that before? But it doesn't really work. It might, if the road were completely level, but as it is she spends all her time either speeding up or slowing down. I put the idle down again and she goes back to using the accelerator. I try to stop for a while every hour for the rest of the afternoon, but the trouble is I'm off in my mind somewhere and I don't remember to.

"You stop," I say. "Just pull off the road when you get tired. I want you to. Please!" But she won't. She doesn't want to slow me down, she says. We make a lot of stops, though, anyway. We stop early for the night, too, and get off late. For the next week we average thirty-two, maybe thirty-three miles a day. The country's the same boring country it's been all along, only there're some small towns now and the stops aren't quite so far apart. Jenny plays the radio most of the day. Sometimes I look in at her and I'll see tears in her eyes. She looks so sad and tired and depressed, so I bang on the window and smile and make her get out of the car and do the best I can to cheer her up. And then we'll move out again and in a little while her face will get that blank look again and I'll begin to feel dead inside myself: empty—like a hollow tree. I'm just a body then, a goddamn robot, a walking machine. Pete's gone. I made him go. Wouldn't let him rest, wouldn't slow down for him. He wanted to finish, too, for John. Two Kunst brothers. I mustn't do that to Jenny.

"I know we just stopped," I say. "I want to check the wagon. You get out and run around." She walks up the road a ways but then she turns and comes back and gets into the car. I slide in next to her and pull her close to me. "Poor baby," I say. "You're very good." She cries quietly and I hold her tight and smooth the damp hair back from her forehead with my hand and whisper to her how much I love her. "Poor kid. I never should have brought her on this. Dummy! Fool! Dork! What do you want to do, drive her away from you?" After a while she's better. That strained, glazed look is gone.

"Oh, Dave," she says, and her eyes have such love in them. "I look over at you sometimes, and you know, you're actually smiling? I feel so miserable when you're the one who's . . ." I hug her to me tighter than ever.

"Don't bother with me," I tell her. "I'm used to this crazy walk." We sit there in the car for maybe half an hour. "We'll stop at a truck stop tonight: get a shower, have a good meal, sleep in an air-conditioned room." I get excited just at the thought of it. We've been on the road for ten days without any kind of real break. No wonder she's exhausted. She looks so happy when I say this that I kick myself again for being such a fool. Our schedule? We're already so hopelessly behind we might as well relax.

"I could ask for another two weeks."

"I thought you didn't want to do that?"

"I'll send a telegram in the morning."

"A telegram?"

She laughs. "That way I won't have to explain anything."

It's my old Jenny again. We're sitting in our cool, clean motel room, getting ready for bed.

> She's my wood.
> I'm her fire.
> We burn together
> With desire!

Another week goes by. We've been stopping at truck stops every night or so, so Jenny's spirits are much better. The problem now is the car. All this low-speed driving has carboned up the engine. I should have taken it for a spin every night but it's too late for that, now. Worse than that, the clutch is slipping. We sputter up a long hill one day along the Murrumbidgee River, near the town of Hay in New South Wales, and the car barely makes it to the top. "Willshe Make It." Jenny thinks of that one. We plaster the name up in tape on each side of the car and write "World Walk" on the rear window. A reporter for the biggest paper in Melbourne stops the next day and says he'll run the story on the front page. Jenny's my mule, I tell him. We joke around with him the way John and I used to with all sorts of people up until we got to Turkey. It's funny how little attention we've gotten from the press in Australia.

Three days later the clutch goes. Luckily, there's a garage nearby and we get it fixed, but that holds us back another day and costs Jenny a hundred dollars. We're close, though, now. Off in the distance we can see mountains. The whole next day our excitement builds. The country is turning lush and green. Rolling hills, our first steep grade. All day we climb the high ridges of the Blue Mountains until, at the end, we camp near the summit of four-thousand-foot-high Mount Victoria. Dense forest around us: huge, sweet-smelling eucalyptus trees. Mist hovers below in the valleys. The mountains really do look blue, too—blue and round and soft, like pillows slung together on a bed. It's cool and windy. The dust and dirt are behind us. In a few days, now, the walk across Australia will be over. We will spend some time in Sydney, arrange for the wagon to be transported to the States, and then Jenny and I will drive back to Perth. The flies are gone. To Jenny, in particular, it all looks like the Promised Land.

We're at a scenic overlook standing with our arms around each other and looking down toward Sydney, hazy and soft looking in the distance. A jet rises into the sky off to our left, makes a circle, and heads off toward America. I'll spend a month back in Perth with Jenny, but then I'll have to leave, too; fly to California and start in on the final lap of the walk: the last two thousand miles, Los Angeles to Waseca.

"It's beautiful here, isn't it?" says Jenny. "Oh, I'm so glad it's almost over." I nod in agreement, but all I can think of is what's in store for me when I leave here, when I leave Jenny. Rather than get on that plane, I'd turn right around now and walk straight back to Perth. Gladly!

Two days later I reach down and touch the waters of the Pacific Ocean. It's the fifteenth of June, almost four years exactly since John and I walked out of Waseca.

CHAPTER 18

I BEND DOWN AND TOUCH THE WATER AND EVERYONE CHEERS and I turn and walk back up to where the crowd is, ready to be presented with the key to the city by the mayor of Santa Ana. Pete's holding the mule and as soon as the ceremony's over we move out through the people: me leading, Pete at the tail. The kids, my parents, Jan, Pete's family, and a lot of other people are walking behind us. The mayor's there, too. We're on our way to Pete's house—about two miles. Tomorrow I'll be moving out alone. The kids and Jan will be flying back to Minnesota and Pete has his job to go to. For the third time on the walk I'll be going it alone. But that's OK with me. I've had my fill of people in the last week.

Jan kept following me with her eyes. After a while I couldn't look at her anymore. I'd be sitting in the living room watching TV, really absorbed, you know, in some Western or in one of those new cop shows. Boy, I couldn't believe some of the things that were on TV. She'd come over and sit next to me, right on me, almost. "David," she'd say. She'd say it maybe three or four times until finally I'd answer her. And then when I did, and I'd lost track of what was going on completely, she wouldn't have anything to say to me. She'd just smile and maybe reach over and kiss me and give me that awful, sad look. "How can she be doing this?" I kept thinking. "I haven't even written to her in four years." But she kept after me.

I hardly recognized her at the airport. But then I didn't expect to see her, or anyone else for that matter. I'd written saying that I didn't want any fuss, that I was planning to get going again as fast as I could, and that it would just be an expense and slow things down if they all came out to meet me. But they came, anyway, Jan, too. I thought at least she'd have the brains not to do that. That was my mother's work, I found out. "All we want is for you to try." She kept saying that to me. How was I to tell her that my trying days were over, that I had a new life now, that Jan was to me as much a stranger as anyone I might meet on the walk. What I wanted, I did tell her, was for us to be friends. Just friends. But no one believed me, Jan least of all. The trouble was that my mother had her all primed to think that now I'd be coming back to her. The walk would get all this restlessness out of my system and I'd settle down in Waseca and take up my old job again, and it would be just like it should have been all those years. Bull shit! But what could I say? I couldn't say anything.

It was good seeing the kids again, but it wasn't the same as the last time. They knew, Debra in particular; they knew I was never coming back. So what was I to them, really? A stranger. The man who walked around the world. OK. I understood that. Maybe someday they'd understand, too, and we could be together again and everything would be all right; but this was the way it would have to be for now.

There're people lining the sidewalks and cheering; a TV sound truck's up ahead. "Santa Ana Welcomes the Kunst Brothers"—a four-foot-high banner right across the street. As we walk under that I turn and look back, and for a second it's John there behind me. It's just like it was, too, except bigger and better. That sign Rich put up on the movie theater marquee, the thirty or forty people who walked with us out of Waseca. Boy, were we green. This Willie's about twice as big as Willie the First, but she leads nice and easy. OK, Pete, give her a touch on the ass. Two thousand miles to go, and all of it in the good old U.S.A. Pete yells something and starts to wave, and I see him grinning and I wave my hand, too, as the mayor of Santa Ana decides he's had enough and joins the crowd of spectators on the sidewalk. Well, that's farther than any other mayor walked with us. And a lot more than the mayor of Newport Beach did. UNICEF, a Commie organization! He wouldn't even recognize our existence. "Hit her again!" I yell back. The mule's slowing down already. Boy, I hope I don't have any trouble with this one.

I didn't want to take a mule at all. It was Pete who talked me into it. "Be good for the image," he kept saying. "You start the walk with a mule. You finish the walk with a mule." Well, he was doing the publicity, so I at least had to listen. "Just as long as it doesn't hold me back," I told him. The mule was trained and everything, he said; just sitting at the Canyon Hills Stables waiting for me. The Santa Ana Chamber of Commerce was dying to present her to us. So, what could I do but give her a try. Her name? Guess. That's right. Willie Will Make It.

When we get to Pete's house, there're still a handful of well-wishers with us, but mostly now it's family and friends of Pete and Nancy's. Pete and I take Willie right over to the zoo for the night, which is nearby. He'll drop me off there in the morning on his way to work.

Jan and I aren't talking. In fact, I'm not talking with anybody much except Pete, so I'm glad it's not going to be another family get-together tonight. We're all supposed to go over to the mayor's house later for a barbecue. The night before last was the payoff. I'd taken Jan to a local bar after dinner—the only place I could think of where we'd be alone (but not too alone)—to try to get it through her head that I wasn't coming back to her when the walk was over. Well, she finally understood. She was crying, of course, when we came back. I was walking a little ahead of her, cold as ice. We came in the door and there were Mom and Dad watching TV in the living room, but no one else was around. They were all spending the night at Nancy's parents' house. Mom and Dad were going off themselves to a neighbor's house for the night. It was all my mother's idea, of course. Flowers in the bedroom, the bed turned down. Jan just burst into tears and I turned around and left, ended up spending the night on the beach. The next day you could have cut the atmosphere around there with a knife. Everything was phonied up, of course, by the time of the big send-off. God, if only I could have just told them the truth.

It's the next morning now, and am I glad to be finally on my way. Willie is moving right out and people are waving at me and shouting from their car windows, and it's as if a great dam has burst inside me and I'm riding along on one of the waves. I stop and get a Coke and a hamburger at about eleven, and a bunch of kids gather around the mule and I sign a few autographs. Most of them saw me in the morning on TV. They're nice and polite and quiet. I tell them a little about the walk and they listen and ask questions afterward. I tell them all to study hard in school, that they're lucky to be growing up in America. It's a beautiful, warm, typical California summer's day, and I can't help but compare the whole scene to what it was like in India, say. "Can we walk with you?" one of them says as I get ready to go.

"Sure," I say, and off we go. There're about fifteen of them. They walk with me for over a mile. Kids come along on their bicycles and circle around and ask questions and talk. Other kids fall in behind. People stop and wave. It's one long parade, with me at the head of it. I'm the Pied Piper leading everybody out onto the open road and I love it. I wouldn't mind if it went on like this forever.

Later on, about four o'clock, I'm crossing a busy intersection and right in the middle of the street Willie decides to stop. I can't budge her. The light changes, horns honk all over the place, but Willie won't go forward. We're like a boulder in the middle of a river. The heavy, early rush-hour traffic flows around us. When the light changes, I lead her back to the curb. It doesn't do any good to beat her. For the last hour she's barely been moving and I've tried that already. She doesn't want to go on, that's what it is. She wants to go back to her soft life at the Canyon Hills Stables. I lead her around the block, thinking to fool her. At first she goes right along; but as soon as we get back to the intersection, she stops dead. Well, I'm determined this time to at least get her across the street. So when the light turns, I dig my heels in and pull. I should have known better, because what she does is plant her hooves out and pull right back. Since she's the stronger she actually drags me back around the corner, so there's a real traffic snarl this time, made worse by the fact that she's backing into it. A patrol car finally gets to us and maneuvers itself behind the mule. The cop switches on his siren and moves forward, pushing Willie in front of him. Only he doesn't, because the damn mule kicks his headlight out and would have bashed in the whole front of his car if he hadn't backed up so fast.

"Get that mule out of here," he yells at me, and I grab her and lead her back in the direction we've come from and off she goes, running. I can barely keep hold of her. Willie Will Make It. Sure. She doesn't even make it out of the city limits.

I call the stables to come get her. I'm sure not going to walk thirteen miles out of my way for her sake. And while I'm waiting for them a man comes over and we start talking. I'm so mad with the dumb mule that when he says he'd like to make me a present of his German shepherd dog, I tell him right off that I don't want anything to do with any damn animals. He goes to his car and brings him over, though, and he is a beautiful dog, I've got to admit. I'm standing there looking at him, and I think of Drifter I and Drifter II and of that shepherd we ran into in India; and I say to myself maybe I should have some sort of mascot. "I'm going to have to put him away," the guy says. "I'll give you one hundred dollars if you'll take him." Well, that decides me, all right. The last thing I want to do is go back to Pete's for the night, but I call him up, and he comes to get me and the dog. That night Mom makes a special blanket for my new companion with "Drifter III" and "World Walk" sewn onto it.

So, the next morning Pete drives me back to the intersection where the mule quit, and I start out again. I'm only twenty miles from where I left two days ago, but this time I'm going to move along at my own pace. A dog sure's not going to slow me down. By the middle of the afternoon, though, I notice that Drifter's limping. Worse and worse. I sit down on a curb, finally, and examine his paws. He's walked the pads right off. What is this, a conspiracy? Am I fated to spend the rest of my life in Orange County? The poor dog can't even put his paw down. All four paws are bleeding. I could carry him, I suppose, but I don't even know the man's name and I'm not sure I could find his house. And what good would it do if I did find him? I don't know what to do. I can't just leave Drifter beside the road. Finally, I pick him up in my arms, carry him over to a nearby gas station, set him down on the grass, and go get myself a Coke. The guy that owns the station comes over and suggests I call the SPCA, but I'm scared to because Drifter's in such bad shape and they might blame that on me. I haven't forgotten that woman who almost got us arrested outside of Philadelphia because Willie had a couple of little harness sores. No sir, I don't want to mess with those people. We sit there in the grass for about half an hour, Drifter whining and licking his paws, and I'm just about to see if I can find a cop who'll help me out when up drives a guy in a station wagon, and it turns out he raises shepherds and would be happy to take

Drifter. What a relief! He offers to put me up for the night, too. I'm pretty bushed, so I thank him and we drive off to his house, and I say to myself, "So tomorrow you really begin the walk. In three days you've made thirty-two miles, but tomorrow you'll step out the way you wanted to all along. Forty miles a day from then on." August first. That's a better day to begin, anyway.

CHAPTER 19

IT'S THE NEXT MORNING AND PERFECT WALKING WEATHER, AND I'm moving fast along Route 91 toward San Bernardino and the mountains. Finally, my luck seems to have changed. I'm walking against the traffic, so people won't think I'm a hitchhiker. I've got on my wide-brimmed white hat that I've had since Afghanistan and a blue knapsack with "World Walk" written on it. I'm waving at the cars and humming along and feeling great. I love walking along this big highway: all the trucks and cars zooming God knows where, all this activity. After all those desolate places I've been in, I love this push-button world, this California hustle and bustle, all the drive-ins, all the gas stations, all the everything. Push, push, run, run. Yowwee! I'm going to walk forty miles a day! No rests. No stops. Just chalk up forty miles a day and pretty soon I'll be in Waseca. The walk'll be over. I'll have done what I set out to do. And then I can fly back to Jenny.

As I walk along a million things go through my mind. It's the first time on the walk I've been completely by myself with no mule, no people hassling me, no problems. I've never felt so free in all my life. I don't know where I'll end up for the night, but who cares? Probably I'll just throw my sleeping bag down under a tree somewhere. No mule to water and feed and stake out and worry about. No one's going to mug me, I've got nothing anybody'd want to steal, and nobody'd rough up a single guy unless he's looking for trouble. I'll stop someplace and get a meal, and then I'll just push on till I feel like sleeping. Maybe I'll camp out in the mountains, up where there's a view. Maybe I'll walk all night. I don't know what I'll do, but whatever it is, it'll be what I want so that'll make it good. The walk's so easy now, I almost feel guilty about it.

"Dearest Jenny . . ." I start writing the letter in my mind. Every day, we told each other. Even if it's just a few lines. This past week there hasn't been much time, that's for sure. "Fair Dinkum," I hear her say, in that soft, sexy, Aussie voice of hers, and I laugh out loud. It's as if she's walking along with me. It really is.

All afternoon it's hot as hell—in the hundreds, with no breeze—but it's a dry heat, so I don't mind it too much. I get thirsty, that's all; but then anytime I feel like it, I stop for a Coke or a drink of water. That's one of the good things about walking through a populated area; there're so many choices. I stop more times than I have to, actually, because I know that in a few days I'll be out in the desert where gas stations are few and far between and there's nothing much else. So I soak up the air conditioning and pack away the liquids. What I really need is a hump. Like a camel. But you can't have everything, can you? About sunset, just outside of San Bernardino, I see a highway patrol station and decide to stop and tell them what I'm doing, and maybe get some tips on what's up ahead. The captain hasn't heard of me or the walk, so it's good that I do stop. We're studying a map and he's telling me to stay off the interstate as much as possible. When I tell him I've been walking with a mule, he shakes his head and says that there would have been no way he could have allowed us on the interstate. There's no point in talking about it with him, either, I can tell. I don't have the mule so it doesn't matter, but I can't help thinking that *if* I had the mule I'd be in trouble because there are a lot of long stretches where all there is is the interstate, and that makes me mad. I'd have had to go cross-country and that would cut my speed in half and play hell with my legs and feet. He asks me where I'm planning to sleep that night, and when I tell him I don't know, he says I can sleep there if I want to. Not in the station. In the garage. Regulations. I don't ask him to show me the regulations or anything. I know this kind of guy. I just take the garage. I've walked thirty-six miles today, and that's close enough to forty, I decide, for my first big day on the road.

By six the next morning I'm walking again, and up until eleven it's cool and beautiful in the mountains. It's high desert all around me, but off in the distance to my left is snow-covered Mount Baldy, a winter sports area, the signs say. Probably people skiing there right now. I pass advertisements for Lake Silverwood, which is way off to my right: pictures of people water-skiing, on vacation, having a ball. It's baking hot and for a minute I let myself imagine that Jenny and I are on vacation—swimming, playing in the lake together. Suddenly, I feel empty and sick with longing. Jenny's half a world away. How long till I'll be with her again? But I push these thoughts out of my mind and look around me for something to latch onto. There are buzzards circling in the updrafts. No. That's not the mood. A silver trickle of stream comes in and out of view on my left, and I move in on that in my mind: clear water, the torpedo shapes of trout hovering in the current. A wave bounces down through the riffles, smooths out across the gravel, stretches its length out over a pool, then builds up again in the rapids. Over the rocks— K-smash! Foam, yellow-white, like the head on a mug of beer, bounces around the smooth side of a ledge of rock. It sits there in the eddy, bubbles bursting, trembling slightly in the wind.

Cactus now. Dry, barren mountains all around. Noon. I cross over the California Aqueduct. Would have been a great place to give a mule a drink, I think, as I lie there soaking my head: drinking, drinking. The sun bakes down and about the middle of the afternoon I begin to go through my old counting routine, just to keep my mind off my thirst. I've got only one canteen and that's not going to be enough, I can see. I'm carrying too much weight, also. My knapsack isn't more than twenty pounds, but it digs into my shoulders and holds me down as if it weighs fifty. There's nothing around now at all: no houses or gas stations or anything. It's beginning to look and feel like Afghanistan or eastern Turkey or Iran. I think of Pete and what he said about the mule. It wasn't just the image. I was going to need her to carry water. The Mojave Desert is up ahead, Death Valley. I'd laughed at him. "This is America!" I'd said. "No sweat."

It's about eight o'clock. There's a lingering, fiery-red sunset, and way off down the road I can pick out what must be Victorville. I'm so damn thirsty I can't think about anything except where I'm going to get my first drink. My eyes are glued to a Shell station sign that I've been walking toward for hours, it seems, without getting any closer. Can it be a mirage? The sign undulates, shimmies in the heat, seems to get suddenly larger, then diminishes again. I'm walking on a treadmill. Then, all at once, it's there just in front of me. Hugging the Coke machine, I pump in one quarter, then another. I don't even taste the first Coke, it goes down so fast; the second either. I drink four Cokes and it's only when I ask the man running the place for change that I realize how swollen my tongue is and how much it hurts. There's a motel across the street and I check into it without really considering that I can't afford it. A shower, a bed, air conditioning, ice water. Forty-three miles I've walked today and I have to admit I'm bushed. It's been six weeks, I know, but that's not really any excuse. I guess I'm still just a spoiled American, after all.

I'm on the road the next morning before the sun's up. I pull my windbreaker out of the pack and put it on. It's real desert country now, all right. The light seems to pour in over the mountain, filling the flat, dry lake bottoms with a bluish light that gradually takes on substance as I walk through it toward the ever-brightening light in the east. Suddenly, the air is shattered by the fiery-red eye of the rising sun. The stillness is gone. A wind blows down from the mountains, and my eyes contract as around me I begin to feel the heat rise. I pull the rim of my hat down over my eyes and step right out, knowing that I have less than an hour before it will be so hot I will have trouble breathing.

By ten-thirty I've drunk all the water in my canteen, my tongue's sticking to the roof of my mouth, and twice now I've felt so dizzy that I've had to stop and rest. Fortunately, I come to an overpass. Stepping into its shadow is like walking into a cool pool of water. I collapse up against the concrete side. There's a breeze. In a few moments I'm feeling all right again, though still aching with thirst. I know I can't go out into this sun until I get some more water, though. I try to remember how much it is the body loses in an area like this. A quart an hour? Can that be right? I think of Afghanistan and all those water jugs we carried. Five, or was it eight, 5-gallon containers? Plus that 3-gallon insulated jug. Didn't we figure a gallon apiece per day? Or was it two gallons? The blessed relief of stopping in that heat—the sound of Willie sucking strawlike at her brimming pailful. Those Afghan soldiers begging us for water because their stupid government sent them out without any. Now I know what they felt like. The putrid slime on the inside of that goat bag that gave us dysentery. "You were lucky you didn't die of dehydration." That's what Dr. Moede said to us when we got to Kabul. Six days we staggered behind that wagon in the baking heat. Up all night. Pull your pants down anyplace. Willie finding her own way.

I curse myself for a fool. One miserable canteen. No salt pills. All across those other deserts we had the wagon. That's why we had it. So we could carry the water we needed, so we'd have some place to get into out of the sun. Just that much shade. The difference between getting your brains scrambled and making it. We'd always stop and sleep during the worst heat of the day. "But this is different," I say to myself. "There's a steady stream of cars rushing by. I can flag one down; get some water. It's not as if there weren't people around to help me." Facing the oncoming cars, I shake my upturned, empty canteen up and down. I point at it with my left hand as if I were in a commercial, and grin as broadly as I can. "Look, folks! A guy without water! In the great state of California, on the interstate, sixty cars passing him every minute, a man in a cowboy hat is dying of thirst. Hey, why don't you stop and help him? Yeah, give him some water. You don't have any? Just stop and see what he wants, then. He's a crazy? What do you mean? How could he hurt you? What's he doing way out here? That's his business. If you really want to know, he's walking around the world. Yeah, the world!" I sit down and lean up against the concrete, panting a little, furious. I can even see their faces; straight ahead they look, their windows rolled up to keep in the air conditioning. They're not going to stop for anything.

I hang around the overpass for three hours. Up right under the bridge there's a flat place to lie down, and after maybe an hour of trying to stop someone, I crawl up there to rest. If worse comes to worst, I can stay here until later afternoon and then walk on until I find water. The town of Barstow's about ten miles ahead. Even if I have to wait till the sun is almost down, I can probably walk that all right. But I can't believe these drivers. Would they stop if I were bleeding to death? And where are the cops? I go back to the side of the road to try some more.

It's almost three o'clock and I'm just standing there when a pickup truck swerves off the road beside me and backs up to where I am. "What ya doin', fella?" It's a curly-haired young guy driving by himself.

"You got any water?" I say.

He looks at me as if I'm crazy, but he says, "Yeah. I got

a whole lot of it." He gets out and jumps up into the back of his truck, and there's a big, five-gallon water cooler. He fills my canteen for me—three times to be exact, and when I tell him how I got here and everything, he just shakes his head. He's a contractor, it turns out, on his way home. He carries water in his truck for his crew. But who else would have it? he asks me. Nobody. And you don't stop because you might not get going again yourself. No one stops out here. Who the hell would ever be out here to stop for? Why he's stopped, he doesn't know. He's still shaking his head when he leaves me. He's filled me up to the chin, though, and with my canteen full too I figure by four o'clock it's safe to push out for Barstow. I walk slower than usual, so as not to get too heated up, though as soon as I step out from under the overpass I can feel the sweat start to pour out. By seven o'clock I'm there. Out of water. Thirsty as hell. But there. And then as I come walking in I hear a siren and there's a highway patrolman flagging me down. What's he want, anyway? When he gets out of the car I see he's got his gun drawn. What is all this? And then I'm spread-eagled against the side of his blistering patrol car and he's frisking me. "OK," he says. "You can go." He drives off without so much as a word of explanation, but up there ahead of me is a gas station with a Coke machine and that's more important to me at the moment than anything else. I'm standing there drinking down my second Coke, still sort of bleary-eyed from everything, when the same patrolman pulls up right in front of me.

"Hey, I'm sorry," he says. "You're Dave Kunst, aren't you? The guy who's walking around the world?"

"That's right," I say.

"How about coming home with me for dinner?" he says, and gives me a big smile. "Make up for being so rough on you."

"Sure," I say, and jump into the air conditioning. On the way to his house he tells me what the frisk was all about.

Just the week before, two of his buddies had stopped some guy like me. Could they be of any help to him? That was what they had in mind. The guy pulled a machine gun out of his backpack and opened up on them as they got out of their car. One of them was hit in the thigh, but fortunately they got the gun away from the guy before he did any more damage, and brought him in. It gave them quite a scare, though. So Dave—that's the patrolman's name—isn't taking any chances.

"Brother!" I said. "This world is full of crazies. That's for sure."

CHAPTER

20

AFTER DINNER DAVE DROPS ME OFF AT THE GAS STATION. HE'D offered me a place to spend the night, but I told him I was feeling good now and wanted to move on.

"Take this with you," he says. "It might come in handy." It's his old army canteen, filled with lemonade.

"What a nice guy," I think, as I wave good-bye. I should have about six of these, right? Even two, though, weigh me down more than I like.

I can't believe how cold it is, suddenly. I'm standing there in my T-shirt, shivering, fumbling for my windbreaker in the knapsack. I'd forgotten about the desert. As soon as that old sun goes down! Actually, it's perfect walking weather, so I move out at a good pace. I wasn't planning to walk far, just twelve miles, to a little place called Yermo, but now I think I may go farther. There're a couple of other small towns up ahead of that. It makes such a difference not to be fighting the heat. I've always liked to walk at night anyway. I guess in my mind I associate it with peace and quiet.

The walking turns out to be terrible, though. Not only is the traffic heavier than during the day, every car, every truck that comes along hits me right in the eyes with its lights. Of course, I could go over to the other side, but I don't want to be mistaken for a hitchhiker—particularly after that story Dave told me. Also, I figure that people will see me better if I'm facing them; and they'll remember that they did, some of them, so if I ever need proof that I was here, walking, it'll be there. I haven't been able to find any mayors to sign my scroll since I left Santa Ana. Maybe they're all off fishing, I don't know. The lights force me to look down, yet I can't see the ground. I trip on a million things, stumble over rocks, pop bottles. I'm afraid I'm going to twist an ankle or something. If I lift my eyes for a second, though, I'm blinded. I have to stop dead, then, or I might walk out into the traffic by mistake. So, all in all, I'm not making much speed. "It's so silly," I keep thinking. "Here I am in America, and I'm having all this trouble." When I finally get to Yermo, at about midnight, I'm more than ready to quit. In fact I turn in at the first gas station and practically beg the guy there to let me sleep in the back. He takes me for a bum, of course, but I don't care. That concrete floor, oily and filthy as it is, feels as good to me tonight as any bed ever could.

The next day I take two Pepsis with me as well as the filled canteens, but the added weight bothers me so much that I drink both Pepsis way before I actually need them. By noon my water's gone, too; but I'm lucky and come to a lone gas station before I get dizzy. When I leave I give the guy there my four-pound tent and poncho, plus a bunch of cardboard-backed pictures of the walk, but even then the pack's too heavy for my liking. All afternoon I trudge along in heat that is well over one hundred degrees, past dried-up lakes with bottoms that look like giant honeycombs or vast stretches of coral. The mountains are closer now, hemming in the air, it seems. Whirlwinds of sand cross the road in front of me, or race directly at me, forcing me to the ground. Grit in my eyes, stuffing up my nostrils. The water in my canteen is hot now, but how good it tastes. Hunched over on the side of the road, I choke on gas fumes as well as dust. The traffic is heavy and continuous, filling the desolation with noise and stink. Night comes and I am still out in the middle of nowhere. My throat is raw, caked with dirt, it seems. I have been out of water since the middle of the afternoon. I keep walking because I don't know what else to do. If I keep walking I'm bound to run into something after a while, I tell myself. And then suddenly, around a turn, there in front of me is, of all things, a Stuckey's. Water—pure, wonderful water, a men's room with a shower. It's like being dropped into paradise.

My shoulders have raw chafe marks on them from my pack straps, so before I go the next morning I decide to junk just about everything I've been carrying. The owners of the restaurant, Sam and Karen Martin, have given me free room and board, so this is also a way of doing something for them. I present them with my sleeping bag, my extra shirts and shorts, my sweat shirt, everything, that is, but my nylon Windbreaker, my newspaper clippings, and my scroll. Boy, what a difference that makes! I can sling my pack over one shoulder now. The only thing I regret is that I carried all that stuff as far as I did. Before I leave, Sam warns me about the rattlesnakes. They have a habit of crawling up onto the road at night. They like the warmth. He arms me with a flashlight and cautions me about sleeping on the ground. He tries to persuade me to keep my sweat shirt, but the loss of weight is more important to me at the moment than anything else. "And don't take your hat off, even for a minute," he says. "You won't hit the real desert till tomorrow. Out there we find 'em dead in their cars all the time." By late afternoon I should be in Baker, just south of Death Valley. He advises me to do most of my walking from there on at night. It's eighty-six miles from Baker to Las Vegas. I don't want to think past there for the moment. That'll be my first real rest stop: four days and nights under the big lights is the way I have it planned. We don't discuss my water supply. I sort of let on to them that that's why I had to get rid of all that weight.

I'm lucky. It gets cloudy shortly after I start out and even rains a little. I reach Baker in pretty good shape at about four o'clock, so after a rest and a good steak in an air-conditioned restaurant, I push on. I don't know how anything could be any more desertlike than the land around here. There were a few places in Afghanistan where suddenly there would be a stretch of low dunes, but desert in most of the world means just very dry, stony country with scattered brush growing and maybe cactus or some kind of low bushes. It's hardly ever just sand. As I leave Baker, the valley widens. The mountains on either side of me are taller, more ominous-looking than they have been, and right now a flamingo pink. As I walk toward them they turn blood red, then dim out into violets and purples, and finally smudge into a soft, velvety black. Immediately the temperature drops: twenty, thirty, sixty degrees? Soon I'm shivering in my windbreaker, walking fast so as not to shiver even more. How cold is it? I have no way of knowing, really.

There is the same steady stream of traffic, mostly large trucks. I'm glad that the shoulder is paved and relatively free of obstacles. I shine my flashlight ahead of me, but it does no good. The oncoming lights obscure everything in glare. Only when there is a lull does the shoulder emerge shadowy in the darkness. Then, for a moment, my flashlight shows up what lies ahead. It glitters on the broken glass, holds in silhouette the curled-up shape of a spun-off tread. I shine it off into the desert. Nothing but an occasional can, bottle, hubcap scattered in the low scrub. Then, for a second, the glint of a pair of eyes. I think of what Sam told me about snakes. Immediately, I see one lying just in front of me like a carelessly flung length of rope. I freeze, move cautiously backward a few steps, my heart beating like a bird's. Blinded, I stop, my eyes shut, head buried, waiting for sight to return. A truck roars by me only feet away. Will the snake move? Will it come toward me? Is it right now preparing to strike. Three trucks go by. A pause in the traffic. I open my eyes. I shine the light in front of me and what do I see? A section of radiator hose! Still, as I walk on I imagine the snakes crawling up out of the desert around me, thousands of them, searching out the still-warm pavement down which I must walk. Once, in a ditch in India, pissing, I stared, hypnotized, as the wrist-thick body of some large snake arched over my foot, gliding from one hole into another. It seemed to take hours. Only when the whole snake had disappeared, when I saw its tail flicker, did I breathe again. I decide to cross the road and walk with the traffic.

What a difference. The lights behind me are my lights now. I curse myself for not having crossed sooner. And right then I see my first real snake. It lies not ten feet from me: still, and long, and amazingly fat; it is stretched a good foot up onto the road. It must be eight feet long. Just as I'm wondering how I'm going to get around it, a passing truck veers over a little and splatters the snake's head to nothing. The body curls up into the air like a whip, falling in a tangled sprawl. I walk by, my eyes glued to its still-twitching form, wondering how many others I may have passed or tripped over without knowing it. A cold shudder runs through me when I think of the other night, walking to Yermo. They must have been all around.

It's pleasant walking, now, for a change. The stars are out, and there's a sliver of moon. It would be beautiful, awesome even, like at times in the Nullarbor, if it weren't for the traffic. We didn't walk much at night in the Nullarbor, but we used to sit up late, outside, just enjoying the silence. I'd walk off into the desert, lie down on my back, and let it soak into me. It was so quiet: no wind, no animal noises. Maybe Pete would cough, or Willie give out a low bray, but otherwise the silence was so absolute that it seemed something I could touch, as if it were made of metal, say—like the inside of a church bell, or the barrel of an old cannon. The only sound was the sound of my own breath, and gradually that slowed down, all but stopped, stopped with the Nullarbor at the point of silence. How long I'd be there, I'd have no idea. Sometimes it turned out to be only a few minutes, sometimes hours. There was no sense of time elapsing. That was the thing. It was more as though time had stopped, paused for a while, and there I was—stopped, too, held within it; just there—waiting.

I hear an unfamiliar noise, a very dry sort of clicking sound, and instinctively I freeze. There, not six feet away, a rattlesnake is coiled, ready to strike, looking right at my extended right foot. I back up four paces. One more step and it would have been too late. And I don't even have a razor blade with me. My eyes are riveted to the snake's. I do not blink. The rattling stops. The head lowers. When I can no longer hear my own heart beating and the strength has come back into my knees, I cautiously move forward again, circling well down into the ditch so as not to disturb my snakey friend. I watch my feet as well as the snake. Who knows how many more there are around? There were no snakes in the Nullarbor. In Turkey, Iran, Afghanistan, only a few. There were plenty of snakes in India, but they were considered a delicacy and we seldom saw one in the wild. No place in the world have there been snakes like there are here. "So pay attention!" I say to myself as I walk on. But it is just that that I do not want to do. Pay attention and the pleasure of the walk is gone.

Toward midnight I find a rest area where there are half a dozen picnic tables. I scout out the place for snakes, examining the trash bin with my flashlight. Then I lie down on one of the tables, and in no time I am asleep. A few hours later, though, my own shivering wakes me and I get up and walk some more, until about 4 A.M. I reach a truck stop where the manager lets me sleep in his camper. The next morning, Lowell, that's the guy's name, brings me breakfast and asks me all sorts of questions about the walk. "The desert's not so very different up ahead," he says, "a long climb up through a pass this side of the Nevada line." He's lived around here all his life, he says. Wouldn't live any place else. People! It's just like it was in all those other countries. The worst places are always up ahead, or back where you just came from, or somewhere else entirely. They're never where you happen to be at the moment. The truth is that what makes a place good or bad is largely a matter of chance. It's cloudy at the time, so the sun doesn't kill you, or there's a flood, or you get bitten by a snake. Fate/chance, I call it. What's in the cards.

The road up to the pass is the worst walking yet. If it weren't for the fact that I get some water at a little crossroads place called Valley Wells—just like some truck stop in the middle of the outback—I'd have passed out from the heat for sure. As it is, I collapse for two hours on a picnic table, in a shady spot, just before reaching the top of the pass at six o'clock. I get to the border as night falls, but after having something to eat I push on. I want to be close enough to make it into Vegas the next night. Fourteen miles into Nevada I find a rest stop and spend the night huddled on a picnic table. I stuff my jacket with newspaper this time, so I don't freeze to death; but it's still not one of my more comfortable nights. All the next day I walk in a sort of trance, stopping at every truck stop, under every overpass. Luckily, there are a lot of them. If there hadn't been, I'd be lying out there now, along with the cows and the horses that don't make it to the water holes. In one place the temperature's 118 degrees.

As the sun starts to go down, though, I see the buildings

of Las Vegas clustered together way out there in the middle of the desert in front of me. I pick up speed a little, a final burst of energy. And then, as I walk past the first hot spot, all at once, as if it were a special welcome just for me, all the lights go on. Wow! What a sight! Like a bomb exploding. Like a million firecrackers. It's fun city come to life for the night. It makes Times Square in New York City look like a funeral parlor. The Dunes, Caesar's Palace. There are neon signs as high as apartment houses. And gorgeous women getting out of chauffeured limousines. All that silver glitter! All those furs! It takes me almost two hours to get to the police station at the other end of town, but I'm so hopped up it's as if I'd flown there. The chamber of commerce has done its stuff this time, all right, and there's a deluxe room waiting for me at the Western Hotel. Even though it's two in the morning by then, I call Pete.

"They want you to appear on the *Tomorrow Show,* Friday," he says. It's early Wednesday now.

"How can I do that?" I say.

"Fly to Burbank Thursday afternoon, dummy. I'll meet you at the Howard Johnson's. I'll bring your clothes."

"Who's going to pay for all that?" I say.

"You get expenses plus a hundred dollars," says Pete.

"See you there, old buddy!" I yell it into the phone, hang up, and call room service for a drink. Later, when I'm sitting on the balcony, looking out at the lights and the darkness of the desert beyond, I think of John and how much he would have loved this. I remember the time we were on the *Mike Douglas Show* in Philadelphia, and what a ball John had with it all. Don Rickles was there, and when Mike Douglas asked John what we were going to use for a mule when we got to Portugal, John said, "We'll have to get a Portuguese mule, I guess, unless Don Rickles here would like to come along instead." Rickles looked as if someone had just kicked him in the face; only for a second, of course, but I saw it. I guess John's probably the only guy in the world who ever ribbed Don Rickles and got away with it. He didn't say a word. Nobody did. Mike Douglas didn't know what to do, either. He just gave a feeble sort of laugh and went on to something else.

CHAPTER 21

"HOLD IT RIGHT THERE OR I'LL BLOW YOUR HEAD OFF!"

My body wants to run, but I tell it to shut up, for Christ's sake, and I stretch my arms as high as they'll go and try to put an innocent-looking smile on my face. There're two cars, one behind, one in front of me; two men, each with a pistol. I shut my eyes so the glare of the spotlights won't blind me. One of the men comes over and searches me for weapons. Then he picks up my knapsack and fumbles through it.

"Hey," he yells. "It's not him." To me he says, "You can put your hands down, fella. We're looking for somebody else." He runs back to his car and the two back out onto the road and go screeching down Route 15 in the direction I've come from. Well, it's no effort now to walk the three miles into Cedar City, even though it's three-something in the morning and I've put in almost fifty miles today. My adrenaline's pumped up so high I can't even relax when I get there.

The guy at the all-night restaurant doesn't know what it was about, so after I eat something I walk over to the police station to find out—and to see if I can get the mayor to sign my scroll. The explanation is simple enough. A guy robbed a gas station a few hours before, hit the station manager on the head, and ran off into the night. When a truck driver called in that someone matching his description was walking along the highway, they naturally thought I was their man.

"Lucky you didn't run," the sergeant said. "They'd have gunned you down for sure."

So much for my welcome into Utah. I almost call Clark to tell him about it; just to shake him up a little for the image he'd painted for me of the state trooper. Clark is the police chief of a little place called Mesquite, just across the border into Nevada. He brought me out some water and sandwiches as I was walking into town and took care of me when I got there. I had more fun with him in that little burg than I did the whole time I was in Las Vegas.

It wasn't just what he did, it was how he did it. He made me feel so good about everything. Kept telling me how great it was for the country that I was doing this walk. He really built me up, I can tell you. Usually when people ask me about the walk, I just tell them where I've been and leave it at that. With Clark, though, right away we were talking as if we'd known each other all our lives. "You and I, we see down the barrel of the same rifle." That's how he put it and that's about right. Clark and I sat up till three in the morning and I told him all about John and Pete and how I felt about all sorts of things. He had to laugh when I gave him my views on marriage and foreign aid and what keeps all those little people back in places like Waseca from getting off their barstools and doing something with their lives. I even told him about Jenny and what I planned to do when the walk was over. If I ever see Clark again we'll probably just pick up where we left off, too.

So what would he say if I called him? He'd probably just laugh. "Lucky those boys didn't blow your head off." That's probably what he'd say. And I already know that.

I guess it's because of Clark, though, that I look up Joe Benson. Joe Benson's a highway patrolman who owns a motel in Beaver, and I want to get some information on the cut-off across the mountains to Route 70, the main road to Denver, but he's not there. His wife tells me that if I want to go on, he'll catch up with me in the morning, but I don't put much stock in that; so I'm really surprised when I hear my name being called on the bullhorn.

"You hungry?" It's Joe Benson in his patrol car with juice and coffee and a big plate of steak and eggs for me. "Didn't figure you'd get this far," he says. It's 7 A.M. and I've been walking half an hour. He leaves me some sandwiches, too. "There's plenty of water in that gorge," he says, "but not much to fill your belly with." He tells me how to get over to Route 70, figures I should make it to a little place called Richfield by dusk. "I'll call and tell 'em to send out the parade," he says, and then he waves and drives off. He's got sixteen miles to go before he gets back to Beaver, and I'll bet he hasn't had his own breakfast yet. So, maybe they breed 'em right around here, after all. Who knows?

I wouldn't have gotten across the desert between Salina and Green River without the help of Tom Shane, that's for sure. Salina's on the other side of the mountains, and from there to Green River is one hundred miles of nothing at all, not even a gas station. He brought me back to his house at the end of my first day's walk and dropped me off again in the morning. Forty-two miles each way. How many people would do that for you? Then two days later he comes by in his tow truck and I ride with him all the way to Grand Junction in Colorado (another hundred miles) where he's taking a wreck; and on the way back we set out caches of food and water every twenty miles so I can make it across there.

On August 21 I walk into Colorado, and right away, almost, the country gets mountainous and beautiful. I'm walking into Grand Junction, feeling really good about having made it across all that desert and looking forward to a couple of days of rest, when this patrol car pulls over in front of me.

"Hello, Officer," I say as the patrolman walks over to me, and I give him a big grin.

"How far you plannin' on walkin'?" he says, as if he suspects me of bank robbery.

"Not past Waseca, Minnesota," I say. "That'll about round out the . . ."

"You're not supposed to be walking here at all. This is the freeway."

"But it's the only road there is."

"You can't walk on it," he says. "You can't walk on the freeway in the state of Colorado."

"Well then, how am I supposed to . . ."

"Don't get wise with me, fella. I'm just telling you the law. Take that next exit. You can walk into town on the state road." He turns back toward his car. "I'll be watching to see that you do," he says.

I'm so angry I can't speak. Fatass cop! How the hell does he think I'm going to walk when there *is* nothing around but the freeway? What does he expect me to do? Bushwhack my way through the woods? I walk into Grand Junction on the state road.

The chamber of commerce has a motel room lined up for me and tomorrow I'll be interviewed on TV. Maybe I should lambaste them there. And then I think of a better idea. In the morning I call the governor's office in Minnesota. I phone up Robert T. Smith of the *Minneapolis Tribune,* too. I don't talk to the governor but I get assurances that the Colorado governor's office will be called, and Smith tells me that he personally will make all kinds of stink if they throw me in jail. "Be good publicity, actually," he says, and laughs. But I'm still mad. Imagine! Here I've walked practically all the way around the world; never been told to get off the road before.

The next morning I walk right out on the freeway, just daring the cops to pick me up. I walk on it all day and don't see a sign of a cop. "Good," I say to myself. "Maybe the bastard got fired. Or maybe someone told him who it was he'd jumped on." Funny not seeing any patrolmen around at all, though. A patrol car passes me the next morning, but it doesn't stop. Hey, fella! I almost feel like flagging him down. That night I call up Smith again and he tells me that the governor's office did put some pressure on, but rather than change the law, the highway patrol decided just to look the other way. Victory! Yahoo! Dave Kunst beats the system!

It's perfect walking weather now, even a little chilly at times. The last few nights, though, have been freezing. Why didn't I pick up a sweater or something in Grand Junction? No brains, that's why. Live and not learn. It's ten-thirty at night and I'm sitting on a bench in a picnic area out in the middle of nowhere, freezing to death and exhausted from all this climbing. I can't sleep out here in the open, I know that. So I go over to a little building that turns out to be a toilet, and lie down on the floor. I can't sleep there either, though. It's too cold. I start pulling out paper towels and stuffing them in under my T-shirt, and then I see that there's a hot-air hand dryer, too, so I push that a couple of times, holding my face right into it. That helps a lot. I keep pushing the thing. Must push it ten or fifteen times. I'm leaning up there against the wall, just pushing the knob on the hand dryer, half-asleep by now, and then I notice that the room's quite a bit warmer than before. I keep pushing the dryer until I'm so sleepy I have to lie down, but by then it's warm. There's not much room. I'm lying there with my head under the sink and my feet practically in the urinal, but I go right off to sleep. About two hours later I wake up shivering, so I get up and push the hand dryer about twenty-five times and then lie back down. The next time the whole routine's automatic. Then it's dawn, and I get up for good and start walking.

Boy, it really is cold out. There's frost all around and I can see my breath. Just before noon I walk into Vail. I'm starving. All I had with me was an orange and some peanut butter crackers, but after I have something to eat I contact the local newspaper and go around to see the mayor. Everything's great. The girl on the paper lets me talk on and on about all sorts of things. For some reason the hand dryer in the rest room makes me think of Esber telling us you eat with your right hand and wipe with your left, when we asked him why it was we couldn't buy toilet paper in his country. And that gets me onto the subject of how backward all those countries are and how much money the United States is wasting on foreign aid. She loves all this. She's really interested in the walk and particularly in what I think about everything. She also makes a big thing about my keeping warm by pushing that hand dryer. "What was I going to do, freeze?" I tell her. She's a cute girl. We hit it off right away. If it weren't for Jenny, I might have stayed around and gotten to know her a little better, but as it is I don't even spend the rest of the day in Vail. After I get done with the mayor, I start up toward the pass. I want to get up and over that and down to a place called Copper Mountain before dark.

It's a hard climb, but just about the most beautiful on the whole walk, I'd say. There's snow up on the peaks and it's so clear and bright I can see way off, as far as the Utah desert. The other way, there's the big spine of the Rockies running north and south and, of course, mountains all around. I'm breathing hard from the thin air, but not too hard. That night I stop for dinner at a Mexican restaurant in Copper Mountain. The owner, Manuel, and his wife not only give me a free meal, they put me up in their apartment for the night. There's one guy there who doesn't seem to like the idea that I'm married and have a family. I can't explain it to him, naturally, and he seems to have problems about his own marriage; but he does the best he can to put me in a bad light. Everyone else is great, though. They treat me like a king. The next morning I get a copy of the Vail paper and there I am on the front page. Where's all the stuff we talked about, though? It's just a straight article about where I've been and what I've done. Manuel loves it, though. He tacks it up on the wall and gets me to sign it. When I go they tell me to come back and stay with them anytime; and they really mean it, too, I think. To them I'm a celebrity: the man who walked around the world. When I tell this other creep that he shouldn't worry so much about what other people think, that the main thing is to do what you believe in, Manuel and two other guys yell out something in Spanish and clap their hands. It sure makes me feel good.

Today's a big day for me. I'm going to be walking over the highest mountains on the whole walk. I haven't decided, yet, which route to take. There's the old, long way over Loveland Pass, which is one of the highest passes in the Rockies, almost twelve thousand feet: a hell of a stiff climb. And then there's the freeway, which cuts right through the mountain with the Eisenhower Tunnel and is about half as long. Normally, of course, there'd be no choice. The problem is that it's absolutely forbidden for anyone to walk through the tunnel, which is a mile and three-quarters long. No exceptions are made. Well, of course, that makes me want to go that way all the more. I'd be the first, just the way Pete and I were the first Americans to walk through the Khyber Pass. I could contact the authorities ahead of time and see if they'd give me permission; but my experience is that it's better to play it dumb. The chances are pretty good that they won't let me through, though; and if I walk up to the tunnel and they make me go back, I'll be losing a whole day plus doing the climb twice. I'd better take Loveland Pass, I think.

When I get to the turnoff, though, I keep straight on toward the tunnel. "What the hell," I say to myself. "I've walked all the way around the world almost, and all sorts of exceptions have been made for me, why shouldn't they make one here?" Also, I figure I'll put up such a stink if they don't let me through that to avoid the bad publicity they'll make an exception for me again. After all, I'm not even supposed to be on the freeway, right?

Well, I'm almost up to the entrance of the tunnel when I hear a siren and there's a patrol car flashing at me from the other side of the road and I figure, OK, this is it. I cross over through the heavy traffic, not quite sure in my own mind whether to play it completely dumb or to take the outraged world walker approach. Before I even get to the guy, though, he yells out, "Dave Kunst, right? Just checking on you, that's all," and gives me a big smile. We get talking and I find out he's from Minnesota, too. He's really interested in the walk, wants to help any way he can.

"Can you get me through the tunnel?" I say. He'll try. He's good friends with the tunnel authorities. He'll ride ahead and see what he can do. "If no one's allowed to walk through, they probably won't let you, though," he says. "You know how these things are." "Yeah, I know how they are," I say to myself. "I'll just have to create my own private little demonstration; they're going to end up letting me walk through, though, believe me. I'm not going back around now, that's for sure."

Maybe twenty minutes later my patrolman friend drives up and I see him shaking his head. I grin at him. "They said they'd like to but there's nothing they can do. It's a very strict rule. No one's ever walked through before. Right?"

"That's about it," he says. "How about riding. They said you could ride through with me. Wouldn't even charge you."

"Are they kidding?" I say, and start walking off toward the tunnel. Ten minutes later there it is, a gigantic double eye bored into the side of the mountain. I remember reading about this thing when they built it: how many years it took them, what an engineering feat it was. Above the tunnel the mountainside goes straight up for maybe two thousand feet. "I'm not going over, that's for sure," I say to myself. The superintendent is standing in front of the tunnel entrance, and he tells me, himself, that he can't allow me to walk through.

"Would you look at these, please?" I say in my politest voice and show him some of my letters of introduction: from Thor Heyerdahl, Princess Grace of Monaco, Hubert Humphrey, the governor of Minnesota. He reads the one by Humphrey through twice and I can tell he's impressed.

"It wouldn't be possible for you to ride through?" he says. "No, I didn't think so. Well," he looks at me hard, and I'm waiting for him to say how sorry he is but . . . What I'm going to do, then, is push past him and get arrested if I have to. We'll see when they stop me if they really want to go through all that. "Well," he says, "I guess, considering everything, we can make an exception in your case. Follow me, please." I'm so startled, he has to tell me again. "But I want you to know that this will be the first time that anyone, other than a worker, has walked through the tunnel." I can't believe it. To tell you the truth, I'm a little disappointed. It's too easy. It takes me a minute or so to feel the old elation.

We go into his office and he orders the fans to be turned on full speed. "We don't want you to die of carbon monoxide poisoning, do we? That would be one hell of an anticlimax, I'd say." And we both start to laugh. We talk for ten minutes or so about the walk. He's very interested in it, asks me all kinds of questions. "And you still get blisters?" he says.

"Sure. I've got one now," I tell him.

"OK," he says, and stands up. "Now see how fast you can make it." He picks up his phone and says, "Kunst is starting now. Let the men know." He points me toward the emergency walkway and off I go, practically at a run.

I'm almost blinded by the lights: headlights, arc lights; I have to walk with my head down, can't see much but my own feet. The fumes are terrible, the noise of the traffic deafening. I pass a guard, every so often, standing in a glass-covered booth and smiling out at me, and I'm tempted to knock at the glass and ask to be taken in there and out of all this for a while, but I keep going. It doesn't seem right, somehow, to rest along here. Eighteen minutes I've been walking and I figure pretty soon now I should be getting there, as I'm pushing it out for all I'm worth. I've got a terrific headache from the fumes and even though I'm trying to breathe just through my nose, every breath sends sharp pains across my chest. My eyes are burning. If the end doesn't come soon, I'm going to have to stop. I keep looking up, hoping to see the end of the tunnel, but there're just too many other lights. And then, suddenly, I'm out. A guard pats me on the back and I get off to the side of the road and bend over and start to take deep, long, slow breaths; and the guard comes over and asks me if I'm all right and I nod my head up and down and then I'm lying on the ground, panting for air.

There're three guards around me, then, and one of them starts to give me artificial respiration, but I manage to sit up and stop him from doing that. My chest hurts, that's the thing. I can't take deep breaths. "Just let me sit. I'll be OK," I tell them. And slowly they move off. There's just one guy who keeps looking over at me from the guardhouse at the entrance to the tunnel. It takes about half an hour for me to feel well enough to start walking again, but after I've walked for a while I'm OK. After all, it's all downhill and there's a good wind and I'm getting plenty of fresh air.

About three miles from the tunnel I stop at a scenic overlook, and off to the east and way down below me, I can see Denver buried in its cloud of smog, and the plains beyond. From there on, as far as I can see, the land is flat. "A couple of hundred miles of Colorado," I say to myself, "then all of Nebraska, up through Iowa and into Minnesota, and I'm there."

CHAPTER 22

I HAVE TO GO RIGHT PAST THE RADIO STATION, ANYWAY, SO I say what the hell and walk in. I'm sick of sleeping in gas stations and in the back of trucks, and that means, unless I want to pay, I've got to get the media to help me out. Publicity! People see your picture in the paper or hear about you on the news, they want to help you. It happens every time. The crew jump all over themselves as soon as I tell them who I am. They've seen me on KOA-TV a couple of days before and they act honored, almost, that I've walked into their little operation. Right away Steve, he's the manager, calls up the Western Motel in Brush, the next town from here, and gets me a free room for the night. "Try the Italian place across the street," he says. "He'll do something for you." I have to practically tear myself away. I guess nothing much goes on in Fort Morgan, nothing to write home about, anyway.

That other night in Hudson was the last straw. The only place I could find that was out of the wind was in between two pop coolers. I had to break down and pay for a motel room the next night, I was so beat. No sir, from here on it's strictly comfort.

> Good bed,
> Good meat,
> Good God.
> Let's eat!
>
> And sleep!
>
> Amen!

And you can't always count on the chamber of commerce. In Golden they were terrific. Contacted all the media for me—as far away as Denver. And they came running, too. But in Denver, forget it. Wouldn't do a thing. So, get yourself in print or on the air or, best of all, on the tube, and you'll get freebies galore. You may not want to work for them, but without freebies you can't walk across your own state, to say nothing of around the world, not unless you're a millionaire.

I take Steve's advice and stop at the Italian restaurant in Brush. The owner gives me a delicious meal with all the trimmings, and when I tell him his food is as good as any we had in Italy, he beams and insists I come back for breakfast. The next day I'm walking into Sterling and a TV cameraman stops and grinds off some footage. He asks me where I'm going to spend the night and when I tell him I don't know, he suggests we ride in to the Holiday Inn and persuade the manager to accept a little free publicity in exchange for bed and board. I agree gladly, and everything goes off without a hitch. The guy films me being greeted by the manager and we do a lot of smiling and shaking hands all around, and then we drive back to where I'd stopped and I walk into town. When I get to the Holiday Inn, the assistant manager takes me into the dining room for drinks and dinner and there's the TV cameraman, too, and we end up having pretty much of a ball, with the TV guy getting a lot of my account of the walk down on tape.

The next morning a car stops and a man and his wife who live near Sterling ask if they can put me up for the night. Well, I explain to them that I've got to keep moving. I'll be thirty to thirty-five miles from there by tonight. They look disappointed, they really do, and then the wife says, "What if we send Billy out to pick you up?" Billy's their boy. Well, what can I say to an offer like that? I thank them and say, sure, and that night, about eight, a kid in an old Chevy pulls over and we mark the spot and in I get. Billy's parents are out for the evening, so before he drives me to his house, we stop at a nice restaurant and he buys me dinner. It's warm enough nights to sleep out, now that I'm in the plains, but I'm not about to refuse the comforts of home when they're offered to me. Still just a spoiled American, I guess.

Early on the morning of September 9 I cross into Nebraska and within a few miles I pick up Route 30. That's the road John and I walked on all the way from Chicago Heights to Philadelphia. "Hey, old buddy!" I say. If I don't look behind me I can imagine the two of them walking along there: hear the clomp of Willie's hooves; John's laugh, his voice singing, shouting at me, whistling at some girl, or low and intimate, talking with somebody who's stopped to see how we're doing.

And right here Route 30 is just the way we liked it: a two-lane road with good, paved shoulders, not too much traffic and a friendly little town every eight or ten miles. That bloody interstate. I've been banging down that for so long I'd almost forgotten what it was like to walk on a real road, through nice country; to stop and chat every few miles, pick up a Coke, some ice cream: walk through the country and enjoy the process. Every time Route 30 turned into four lanes, John and I would groan. It's big, open country along here—grazing land. Since Fort Morgan the road's been following the South Platte River. It's wide and slow and muddy looking. Makes me think of the Ganges. Right here, though, in the upper Midwest, is probably the richest, cleanest, and best-off area in the world; and the Ganges goes through the worst, poorest country there is. Just thinking about it makes me feel good to be an American. And lucky. OK, we've got a lot of problems, but we're still the richest, the best, and the most beautiful country in the world. I thought that before I went on the walk and I know it now. There isn't any comparison.

John and I used to rate the countries when we went through them. A hundred years behind America. That was Portugal and Spain. Fifty years for France. Seventy for Italy. Western Yugoslavia was about the same as France: nice farming country. We were surprised. But Bulgaria was so dead we couldn't even measure it. Nothing but police and scared people. We stopped comparing altogether when we got to Turkey. And with Pete and me the subject didn't even come up until we got to Australia, everything was so backward in all those countries we went through. Australia was only ten or fifteen years behind the United States, we decided. But that was before we got into the outback. Imagine the Mojave stretched all the way across to eastern Pennsylvania. That's Australia.

I look into a field and there's one of those huge harvesters, shining red and bright as a fire truck, eating up the wheat. Around it, the fields go on forever, it seems, like the sea. I watch the wind come down and make a path for itself and then disappear, and the heavy heads slowly right themselves until it all looks the same again. I'm sucking in every breath because it all smells so good, even the dust along the road. John's with me, and Jenny, and my mind is going back and forth over the whole walk. I'm flying in that tiny plane in the mountains of the Hindu Kush; then I'm with Jenny in some truck stop in New South Wales, and we're laughing. The clean streets of Singapore. Pete and I just walking around, taking it all in.

"David Kunst?" There's a fellow there behind me, not twenty yards away, a pack on his back, hiking boots. "You really can walk," he says. He comes over and we shake hands. "I didn't believe you could keep up a pace like that when I read about you, so I came out to see for myself." He spends most of his time hiking around, he tells me; says the best he's ever done in a day, in flat country, is thirty-two miles.

"Well, I'm walking on smooth surfaces and I don't carry anything," I say, mostly to make him feel better. I know, though, that I could do better than that. My stride's ten inches longer than his. I'm a natural-born walker, that's all. Could he walk around the world? I doubt it.

That afternoon I pass a sign that reads "Buffalo Bill Ranch: Historical Park." I'm tempted to take some time off to see it. It's only four miles out of my way. But I decide not to because at the last minute I say to myself, "Why? Why do you want to go?" And I can't answer my own question. To see Buffalo Bill? But that's ridiculous. To see where he lived? But it won't be really where he lived. It'll be touristed up. To say I've seen it? Forget it. It's like so many other places we passed on the walk. You can't miss that, someone would tell us. Go through France and not see Paris? Impossible! I'm not a sightseer, that's all. A lot of the famous sights in the world aren't that much, anyway, when you actually see them. There's a lot of dog crap around, say, or there're beggars. Some of the most interesting things we saw were worth looking at not because something used to happen there—like the Roman aqueducts, for instance. They were places or things that would be good to look at anytime. Like the chapel Marieanna took us to in Évora, Portugal.

It was Christmas Day and Marieanna was waiting for us, as planned, when we arrived at the Hotel Planicie. After we had Christmas dinner with her parents, she took us to what she said was a very holy place. It was in the cellar of the chapel of the Monastery of St. Francis. One of the things that threw me off was that I had always connected

Saint Francis with birds, maybe because of this picture that hung in my room when I was a kid—an old-fashioned, brown-and-white tinted picture of a very thin monk with a long robe on, feeding grain or bread or something to some birds. I remember what used to impress me was that the birds weren't just on the ground in front of him. They were sitting on his head and shoulders, and four or five of them were eating out of his hand. I loved that picture. I wish I still had it—although now it might not mean the same thing to me. Anyway, we went down these cellarlike steps and I was all prepared for something nice—nice and holy, like she'd said; maybe something to do with birds. Marieanna crossed herself about a hundred times and then we went through a narrow doorway and she took us over to a wall where there was a prie-dieu with a crucifix over it lit up by two candles. She got down on her knees, while we just stood around, and then she got up and stroked the crucifix with both hands and stepped back and motioned us to come forward. Well, I wasn't about to go through all that. It had been a long time since I'd been an altar boy, so I just sort of came forward and looked more closely at the crucifix. There was something odd about it, but I couldn't figure out what it was at first. Then it sank in. The crucifix was made out of two long bones, what looked like very thin, long leg bones. There were jewels and goldwork on it, too; but all attached to bones. I stepped back fast, believe me. "Saint Francis *spiritus*," Marieanna said, crossing herself and bowing her head. Then I looked down and I saw that the whole prie-dieu was made up of bones. The wall next to it was decorated with bones, also. In fact, as I looked around, I could see that the whole chapel was built out of bones. I moved back toward the door. I couldn't see very well in the weak light, but I thought that even old John was looking a little green.

All this time, Marieanna was wrapped up in praying. Finally, she quit and stood up and came over to us with this sad-looking smile on her face and said, "Is it not beautiful? I think always of the nuns and monks. It is they who make this chapel so special. They did it for me, for my soul, for me to understand and love more deeply."

Well, what could you say? We sort of smiled and nodded and hoped she was getting ready to clear out. No such luck. She went on for another ten minutes at least about when the chapel was built and how long it took and how special it was because it was the only one of its kind anywhere. Well, I could see how there might not be too much demand for this sort of place, but naturally I didn't say that to her.

"Why did it take so long to build?" I said, more to say something than really to find out.

"Because we must wait for them to die," she said, very softly.

"For who to die?"

"For the holy nuns and the brothers."

Then it hit me. "You mean these are human bones?" John and I were out of there fast. Imagine praying inside of a building made out of human bones. I could just picture all the thousands of dead nuns and monks—young ones and old ones and some dying in terrible pain and others wasting away from something or other that no one could figure out because there weren't any good doctors in those days. All of them dying and then lying around somewhere until the flesh rotted away from their bones. I was trembling by then, and not just from the cold. I saw the flesh turn white and then get puffy and black, and finally the little pock marks appeared in it that became holes, and then everything started dripping, as if it were all made up of very dirty, yellow wax. How long would it take before the first bone showed through? The fingers would come clean first, I figured; then the toes. Maybe the nose next. Were there bones in the ear? I reached up to feel my own, but my hand was shaking so I brought it down again. It would take years, and there would be someone whose job it was to look after the bones. The surveyor of the bones. Some monk, probably. Imagine spending your whole life sorting, stacking, carrying human bones. They were very white, so maybe he bleached them out in the sun like sheep's bones when it got to be that time. What did he do with the skulls, I wondered? And then I almost passed out. Those weren't cobblestones we were walking on. They were the smooth, round skulls of human heads—polished and darkened from all the feet, and with little cracks in them. That was sure some place to take a person. And on Christmas Day, too. It was one of the most interesting places we went to on the whole walk, though. If I shut my eyes, I can see it now, feel the smoothness of those old bones against my fingers.

In the past three days I've spoken at a Lions' Club meeting, been on the radio, and given two newspaper interviews. I've been taken out for dinner every night and slept in three different houses. I haven't bought a meal or spent the night outside for more than a week; and the farther I get, the more curious people are, it seems, about me and the walk. In one family's house, some people come in after dinner to meet me. Some of them are shocked by what I have to say, and tell me so, but they're plenty interested, too. No one leaves until it's good and late and they really pump me with questions, not dumb questions either.

"I think everybody goes to heaven, that's what I think. There's no hell. Just heaven."

"Everyone?" says a middle-aged man with a bushy, gray beard, who hasn't said anything all evening. "Even Hitler?"

"Sure," I say. "Especially Hitler."

"And that's justice?" the man says.

"Absolutely!" I see by their faces that no one agrees with me. It's so clear to me what I mean, but I can't seem to explain it. God created the world on a whim. OK? He just set it going. Like a great, big bowling ball. What happens is fate. Even God doesn't have control over that. So what's he going to do, punish some dork when he dies by making him suffer forever? Not on your life. The poor dork didn't know what he was doing, probably. No. He's going to take it easy on him, give him a break.

"Call it mercy, if you want to," the man says. "But there's nothing *just* about it."

"There is, though. You don't understand." I'm getting angry now. "Say a guy comes in here and kills us all. He's sick. He doesn't know what he's doing. Well, we'll all be in heaven afterward." The man with the beard looks disgusted. Most of the others are confused. "And it won't matter," I say.

"It would matter to me," the man says.

"It won't matter, because you'll be in heaven. You can't feel resentment or jealousy or anything else like that there, so you won't care that that man who killed you is there, too. You won't care." That shakes up some people. I hear grunts of disapproval and annoyance. There's a girl in the corner who looks up at this point and smiles at me like crazy, though. The man with the beard just lights up a cigarette and says something to the woman next to him and then they're both smiling.

People start to leave. I'm excited. I'd like the talk to go on longer. After they all go and I'm in bed, I think of other things I could have said. There's no hell, because if God's good he couldn't create such a thing as eternal punishment, could he? It wouldn't make sense. That's why everybody has to go to heaven. Everyone's life here on earth can be improved some, that's for sure. Have you ever met anyone who couldn't have had things a little better? And think of all the poor, miserable people there are in the world. They've had enough hell right here. They sure deserve better after they die. So everyone goes to heaven and that way even those people who have had it pretty good in this life won't have it above anyone else there. It'll be the same for everyone, that's the thing. So it doesn't matter. It really doesn't matter what happens to us here on earth.

Two days later, in Grand Island, I call up the police to let them know I'm walking through, and they give me the message that a reporter from the *Minneapolis Tribune* is looking for me. "Fine," I say. "Send him over." And I tell the sergeant where I'm staying. An hour later he arrives: Warren Wolfe. He brings greetings from Robert T. Smith. "They want me to walk with you for a couple of days," he says. "Find out what you're thinking. What kind of guy is it who walks around the world? That's what I'm supposed to dig for." We laugh. There's nothing I'd rather talk to him about.

CHAPTER 23

Minneapolis Tribune, September 22, 1974, p. 1ff. Aurora, Neb.

AN OLDER AND WISER DAVE KUNST NEARS WASECA

by Warren Wolfe

Dave Kunst, 35, glanced over his shoulder to check for traffic as he walked along Hwy. 34 in eastern Nebraska, his long, bouncy stride carrying him toward his hometown of Waseca, Minn., at the rate of about 4 miles an hour.

"I'm a social deviate, a radical, an oddball, even a little crazy," he explained between anecdotes about his trip. "I don't fit into anybody's pattern and I never will."

An energetic, impulsive, loquacious man, Kunst is only a few hundred miles from the end of a 4½ year, 15,000-mile walk around the world for UNICEF.

"I feel a little uneasy about ending the trip," Kunst admitted. "My home has been where I hang my hat, and I like that feeling. I'm not ending the trip to become domesticated."

He'll walk into Waseca at about 1:30 p.m. Oct. 5 to a celebration led by Sen. Hubert Humphrey and Minneapolis businessman Hal Greenwood, two of his trip sponsors.

The long journey has left its imprint on Kunst. For one thing, he's learned a lot:

"I had no idea there were so many damn dumb foreigners

in this world. And bureaucracies; you think ours is bad, you ought to see what they're like in Asia. And I've learned a lot about myself; I know that I'm not going to waste time doing things I don't like anymore, like trying to make a marriage work when there's nothing to work with."

"Sure, I've changed some since I started the walk," he said, "but I've always been different from most people. It's just become more pronounced during the walk. I'm happier, I don't hold back my punches when I talk to people. I've had a hell of an education."

And he has more confidence.

"I know I can do just about anything I make up my mind to do," he said. "I know I've got a lot of good ideas and I'm a pretty good persuader. People don't always agree with me, but they usually respect my attitudes after I've talked with them."

Kunst is an intent listener as well as a talker. He has strong opinions on everything from marriage to foreign policy, and is quick to challenge someone who disagrees with him:

"I'm not always right . . . well, hell, I'm nearly always right," he said, his weathered face crinkling into one of his frequent grins. "If somebody argues with me, I'll listen. I may even change my mind, but not while he's there. I'd never admit to somebody that he convinced me I was wrong."

He can talk for hours about himself, his attitudes, his experiences, and freely admits that "I've got an ego as big as they come. I love the attention I get from the press and television and from people who've heard about me and stopped to talk."

The walk has been spangled with exciting experiences: meeting Princess Grace of Monaco; trooping into an Italian restaurant, mule in tow, to lunch with author Thor Heyerdahl; visiting with diplomats and military brass; seeing beautiful scenery.

"And, let's face it, I'm not a priest," Kunst said. "I've known a lot of women all over the world, and it didn't always end with a handshake. I'm not bragging, sex is a natural part of life. Going without for 4½ years would be unnatural."

But there have been bad times: crossing deserts that are

cold at night and blazing hot in the day; being stoned by children in Turkey; drinking foul water and getting amoebic dysentery; trying to buy goods in a store when you don't know the language and "even the sign language is different."

The worst was the shooting in Afghanistan. Dave was shot and his brother, John, who started the trip with him $2\frac{1}{2}$ years earlier, was killed.

"I was surprised at how cool I stayed during the shooting," Kunst said. "I was shot, and I yelled at John to play dead. I didn't know then, but he already was dead; he'd been shot through the heart. The Afghan bandits had been following us for a couple of days. They thought we had money because the newspaper said we were collecting money for UNICEF.

"They came up to us and shot John again. Then one of them pulled my watch off. I was thinking. 'Thank God I'm not wearing any rings,' because they'd have cut off my fingers."

The experience and one about seven years ago when Kunst was injured in a traffic accident as head of the county surveying crew "have had the most effect on me of anything in my life," he said.

"I realized, by God, I could die any day. But before I die there are a whole lot of things I want to do. I'm not going to get killed at age 35 or 40 or 50 and have my last thought be, 'If only I'd done those things I wanted to do.'

"No, I'm going to do what I want to do, what I need to do, and this trip around the world is just the start," he said.

After John was killed, another brother, Peter, joined Dave for a year, walking from Afghanistan through Australia.

Walking for UNICEF was John's idea, Kunst said.

"He thought we ought to be doing it for something more than just ourselves. The main reason for the walk, though, was because we just plain wanted to do it. Nobody had ever done it before."

The brothers handed out UNICEF-donation envelopes around the world. UNICEF officials say they'll never know how much was raised by the walk because donations have been sent to various centers around the world and some countries did not keep track of how much came because of the Kunsts.

Minnesota UNICEF officials say $3,484 has been collected at the Minneapolis office. New York officials are trying to find out what they can about how much has been donated worldwide.

Kunst isn't sure how much money the trip has cost him. He withdrew the $3,000 from his county retirement fund and split it with his wife before leaving. He's borrowed a good deal more. His wife, Jan, works in Waseca to support herself and their three children.

Some people in Waseca tried to dissuade him from making the walk, and once he started many doubted he'd finish.

"There are some real good people in Waseca," he said, "and there are some awfully small people there. I was home for three months after the shooting and went down to the bar, and those guys were just the same as they were 10 years ago. They sit there with their fists wrapped around their beers and bitch and moan about their jobs and their wives and their lives, and they escape by going to the bar.

"Well, that's not real life. Those guys don't have the guts to do what's best for them. I respect a guy who says he wouldn't take a trip like mine because he'd miss his wife and kids, but I really hate a guy who says he wouldn't go because his wife wouldn't let him."

In his own case, he said, his wife was not eager for him to leave.

"Let me tell you about my wife," he said. "She's a good mother and she'd make a great wife for a guy who likes to come home every day after work to a nice, quiet meal and a nice, quiet life.

"We haven't seen eye to eye on most things since long before the walk. I asked her for a divorce a couple of years before I left. She didn't want to give me one, and I guess that's OK if she wants it like that. I don't plan on getting married again anyway. But I'm not going to be spending too much time at home."

His first project after returning to Waseca will be to write a book on his experiences. He figures that will take about nine months.

"It's not going to be a travelogue. It'll be more about what I learned and thought as I traveled. If John were alive we'd have had a big argument about this. He wanted to do more of a book about our experiences.

"I'll write the book for whoever will give me a $10,000 advance and some kind of percentage off the sale of the books. Most of the money will go to my wife. I really don't care if it sells or not—don't get me wrong, I'd like to make a million—but I want to write it for myself."

The walk has stretched his imagination about a lot of things, Kunst said: "You've got to come up with unique ways to combat illness, the heat, the cold, lack of water, all those problems you run into when you don't have somebody else handy to bail you out.

"Like the time in Colorado when I was walking at night along the interstate, and it was so damn cold. I stopped at a rest stop for a few hours' sleep, but it was too cold. So I went into the men's lavatory and kept punching that damn button on the electric hand dryer until the place warmed up. Now who else would think of that?"

Kunst walks anywhere from 20 to 60 miles a day "depending on how I feel and how much in a hurry I am." Now he's carrying a light pack, with one change of clothes, a toothbrush, razor, a pocket dictionary (the only item that he's had for the whole trip), and an empty canteen ("It looks good swinging on the back of the pack").

And he still gets blisters.

"I've been trying to save this pair of shoes," he said, "but I guess my feet are more important." He stopped, bent over and carved a hole from the top to make room for two blisters on his toes.

"Let me tell you," he said, waving his shoes, "these are the best damn walking shoes I've ever had. They're made in Minnesota and there's nothing like them in the world." They're his 21st pair of the trip. He also has been supplied clothing by a Minnesota firm since the shooting.

As the end of the walk nears, the book he'll write becomes more and more important.

"That and the car—did I tell you a car dealer in Waseca has given me the use of a car for a year? I think that's pretty damn nice—are what I'm really looking forward to now. The walk is over right now, for all practical purposes. I'm just coasting," he said.

"In the book I want to tell what I think about people, people I like and some I don't," he said. "I want to write about my attitudes, my parents, my family, people I know in Waseca, all the things that have gone together to make me what I am now."

Among his attitudes that he says will be part of the book:

—Marriage: "Generally, I think it's a bad idea. I've talked to too many people who feel trapped by a spouse, who want to do something—even something as simple as staying out late some night—and don't because the wife or husband wouldn't like it."

—The walk: "I'm doing this for myself mainly. I was tired of Waseca, tired of my job, tired of a lot of little people who don't want to think, and tired of my wife. The walk was a perfect way to change all that: I just walked out of town."

—Environmentalists: "I hate most of the ones I've met. They're a bunch of do-gooders who aren't at all realistic. They moralize about pollution and if they had their way they'd shut down half the businesses in America."

—America: "It's the best damn country I've ever seen. It's got the best mountains, the best agriculture, the best businesses, the cleanest rivers and lakes, the best government, the best standards of living, the best food, and the smartest people."

—Foreigners: "I've never seen so many ignorant people in my life. You can really appreciate our education system when you see people in some other countries. They're really stupid. Part of it's lack of education, but part of it's their cultural heritage. They need more self-confidence."

—Foreign policy: "We take a lot of crap from a lot of crummy little countries. We're the most powerful nation on earth and we ought to act like it. That doesn't mean we should tell other people what to do. But we don't need to give aid to countries that don't want to help us."

—Religion: "I've got a fantastic relationship with God. I used to be a pretty strong Catholic, but every church I know about tries to tell you how to act and what to believe. I don't like that. Everybody's relationship with God is personal. You have to find your own religious system, your own values of right and wrong."

—His brother, John: "I think he can hear me. I talk to him and God sometimes when I'm walking down the road. At the end of a 60-mile walk I say, 'John, you son-of-a-bitch, you'd have never made it 60 miles today.' I miss that guy sometimes. I hope he can hear me."

—The future: "Life is a journey, not a destination. Somebody else made that up, but I sure agree with it. I don't know for sure what I'll be doing in about a year, after I write my book, but I've got some plans.

"By the way, did I tell you? I'm going to live to be 140 years old."

I READ THE STORY STRAIGHT THROUGH AS IF IT'S ABOUT SOMEone else, and I have to admit I love it. Finally, someone's written down what I think about things. I have to laugh at some of it: sixty miles. Come on! But mostly he got it right. Makes me sound like a real bastard, but I kind of like that. "A social deviate." So, what's that, some kind of criminal? "Crazy?" Some people would sure say so. I like the way he puts all my ideas together at the end. Makes them sound important. Send a copy to the President. That's what I should do. Take me on as an adviser. I sure never expected Wolfe to come out with something like this. He can't walk, that's for sure, but he certainly can write. I stand there in the drug store in the middle of Omaha, grinning like crazy, reading sections of the article over again to myself. There I am on page one as big as life. Headlines right across the front. It sure makes me feel good.

I tuck the paper into my pack, walk down to the bridge, and cross the wide Missouri into Council Bluffs, Iowa. As soon as I get out of town on Route 191, I'm in a different world. Omaha's the end of the open rangeland of the West, and this is the beginning of the real midwestern farmland: checkerboard fields with a farmhouse in the middle and some little town always on the horizon. It started in Nebraska: that friendly, outgoing feeling; but this is real God's country. There's no richer, friendlier, better place in the whole world than the farmland of Iowa and southern Minnesota. I keep going over the article in my mind. I guess I'm as happy about it as I'm supposed to be about all the celebrations that are coming up for me when I finally walk into Waseca on the fifth of October. Sure, having the free use of a car for a whole year sounds great. The only trouble is I won't be around to enjoy it. As for the fuss the town fathers are planning, what I'm most interested in is finishing the walk, getting it in the record books, and then beating it back to Perth, Australia, and Jenny. All the attention I've been getting recently is going to my head, though. I've got to admit it. To be treated like a hero by the governor of Nebraska! That doesn't happen every day.

I'm walking along the road the next morning feeling great. I've just spent the night with the mayor of Neola: my first official welcoming committee. People are waving from their front porches as I go by, and some walk down to the road with glasses of cider or cookies or something. One old fellow yells out, "What I say is, everyone's got a right to his own opinion." I wave and think, "That guy's read the article, I bet," and for the first time I wonder what sort of reaction it's going to get in Waseca. Just before noon a patrolman stops and tells me my hometown paper's trying to get in touch with me and for me to call them. It's about time they did, I think. I'm looking around for a phone in the next town when three cars pull over in front of me and six reporters and two cameramen get out. By the time they leave, I'm not all that anxious to make the call.

It seems that Waseca has overreacted to what I said in the paper. To say it straight out, the whole town's furious. The thing that got them most worked up, according to one of the reporters, was all that stuff about my love affairs. Well, for Christ's sake! And all I did was admit I hadn't been exactly a monk. The same people who were saying how shocked they were by me were the ones carrying on like crazy themselves. I could assure them of that. "Would you care to name names?" one of them says. Well, I could, of course, but I don't see any reason to do that, so I tell him, "No." What hurts me the most, though, is that Mr. Greenwood has pulled out his support and condemned me in the press. He was our first and biggest backer. To have him think that a few remarks of mine make the whole walk meaningless makes me feel terrible. How can anybody react like that? As if he personally were being betrayed. It didn't surprise me to hear that the free car was off now, and that Carl Swanson and a lot of others had spoken against me and urged that Waseca call off its celebration. I was pretty shocked that there were actually threats of physical violence, though.

"I guess you could say I persuaded them, David," says Oather Troldahl. He's the editor of the *Waseca Journal* and president of the Waseca Chamber of Commerce. He's been one of our best supporters and I'm glad to find out he's taking the whole thing pretty calmly. "We had a meeting this morning and there'll still be a celebration," he assures me. "The emphasis will be on the walk, now, that's all." He's gotten the town worthies to take a "Christian attitude" toward me. How about that? But this is the payoff. In the middle of the meeting he took everyone outside of town to where there's a sign that says "Jesus Loves Me." You see it on your left as you come in from Janesville. The Kiwanis or someone put it up. Well, Troldahl had a local sign painter write underneath it, "And You Too Dave!" Can you imagine? And they loved it. They turned the other cheek. He read me the resolution they adopted when they got back to the meeting: "Let's forgive critical statements about Waseca by Kunst for the greater good of Waseca. Let's support Kunst's wife, Jan, and his brother, John, who was slain by bandits while accompanying his brother on the marathon walk." They vote to support the walk, but not "Kunst's ideals." For the first time in my life, almost, I don't know what to say. The thing is he's expecting me to thank him. He's done his best.

I don't feel like talking to anyone that night, so when I walk into Denison I just get myself a motel room. I'm sitting there watching television when the desk calls. Some newspaperman wants to talk with me. Well, I want to tell him to bug off, but instead I say, "Put him on." He's a columnist for some big paper, it turns out. Wants to walk with me next day. "OK," I say, and we arrange to meet at a restaurant in the morning. No, I don't want to talk to him right now. I want to get back to the TV program I'm watching. He's there when I walk in for breakfast and practically the first thing he says is, "Wolfe made up all that stuff, didn't he?" I tell him, no, I'd said most of those things, maybe not in quite that way. He gives me a knowing look and says, "Publicity stunt, huh?" No, I tell him. Definitely not. "You're fixing to make a lot of money on this, aren't you?" he says. "Tell me." And he leans way forward. "Did you really walk every step of the way?" And then he laughs. Well, he can't keep up with me, so pretty soon he drops out of his own free will and I'm rid of him. I don't tell him much, that's for sure. Who knows what he's likely to make of what I do say?

I'm walking along enjoying being by myself for a change, when who should drive up but my parents. Boy, am I glad to see them. To think of their driving all that way, and on a weekday, too. Dad's got a pretty strict schedule: northwestern section of the state one day, middle-south the next, and so on. He's changed things around, he tells me, so he'll lose only half a day's calls. They'll be going back early in the morning. But they're not here just to say hello. Mom's got something on her mind, I know from the way she's looking. And pretty soon she comes out with it.

"You don't have to live with those people, that's the thing. For you it's different."

"What's the matter?" I say. "Mabel Young tell you what a bad boy you had?"

"Wasn't just Mabel Young," she says, and her eyes get real small and mean looking.

"What'd you want me to do, lie to him?" I say, looking from one to the other.

"Too much honesty can hurt people sometimes," she says. She's mad as a hornet now. "I can think something, but why do I have to tell the world about it?"

"Why not tell the world?" I say. "If it's what you think?"

" 'Cause it might hurt someone," she says. "Don't you ever think of that?"

"Who have I hurt?" I say, but I know from her face who it is she's talking about. "Jan's got to take it, that's all. She and I had that out in California."

"And what about Debra? And the boys?" she says.

"The same with them," I say. "I know it's tough . . ."

"It's not fair," she says.

"There, there," says Dad, and puts his arm on her shoulder. "David's always been pretty outspoken. We've lived with that." Later on, in the motel, while Mom's out of the room, he pulls me aside. "It was just one thing bothered her, mainly; what you said about her." I can't think what he means and tell him so. "That part there where you referred to her as a female dog. Now we know you didn't really mean that. Still, no one likes to see something like that in print." I can't think what he's talking about, so I get out the article and he puts his finger next to the passage. It's right toward the end where I'm talking about John. "John, you son-of-a-bitch . . ." That shuts me up. I can't believe it. They've driven 150 miles to ask me please not to embarrass them in public any further. I tell them I'm sorry if what I say hurts, but if someone asks me something, I have to tell them what I think. Everyone's got a right to his opinion, I hope, and that goes for all of us. They don't say a word. They have to agree with that.

And then I hear my mother say, "It's different where your own family's concerned." Dad breaks the silence. "I guess it gets back to the old saying about the Indian. Before you condemn him, you should walk a mile in his moccasins." I'm sure not very sorry when they leave.

About mid-afternoon the next day, Oather Troldahl and one of his photographers drive up. How am I feeling about what's been decided? he wants to know. I shrug my shoulders. What kind of a question is that? Am I mad at Waseca for how it reacted? Not really, I tell him. There're some good people and some small people in Waseca. Like any place. They overreacted, that's all. My comments were just my personal opinions. I'm not always right. I know that. But right or wrong, I have a right to say what I think and so does everyone. And that's a right I would defend to the death, and I hoped every American felt the same. Am I still planning to walk into Waseca on the fifth? That's what he's really come to find out. "Sure," I say. "I'll be there."

After he leaves, the question of why Waseca reacted the way it did goes over and over in my mind, and I finally come up with this explanation. Ever since the walk started, there have been people against it for one reason or another. At first they said we'd never make it. After John was shot they said, "Hasn't he learned his lesson yet?" All along there were remarks about my leaving my wife, abandoning my kids, quitting my job. Toward the end of the walk, though, these people had to shut up. It looked as if I was going to make it. Kunst was going to be Waseca's biggest name from now on. "What would that do to morality?" these people were thinking. Who knows how many other people would walk out on their wives and families and jobs and use me as an example? There might even be a general walkout. People might decide to do what they wanted to do for a change and not what their families or their friends or their churches or their bosses told them they should do. And that would be bad for society, wouldn't it? Society would crumble away and die if that idea caught on, wouldn't it? To these people I was a devil, a force for evil and destruction. They hated me, I suddenly realized. They envied me and they resented me and they were jealous of me and as soon as they saw their chance, they hit out at me. They'd take the walk away from me if they could. But they can't do that, so they'll do the best they can to spoil it.

Well, I don't care about that. I care about having done what we set out to do. That's all. If there's no one in Waseca but the press, it'll be all right with me. Things are getting out of hand, though, so what I do is I prepare a statement for Troldahl, and anyone else who wants to know what I think. It seems funny to read a statement instead of just speaking what's on my mind, but that's what I'm going to do from now on. This is it:

> I think the whole thing is silly. Waseca overreacted, in my opinion, and I feel a man is entitled to his own opinion. A certain, select group felt threatened by what turned out to be only me, they realized after the initial shock. It makes me wonder . . . But I will walk into Waseca on October 5th to finish what John and I started. The important thing is I did it, just the way I said I would.

About four o'clock my mother drives up. She's in a completely different mood. Some of the Waseca town toughs are headed out this way to rough me up, she says. That's what she's heard and she wants me to know about it, so I'll be ready to defend myself.

"Don't worry about me, Mom," I tell her. The other stuff's all gone. It's as if she and Dad hadn't been out at all. We have a great time together at dinner. It's like the old days. I'll say something about what happened when . . . and she'll add to it and that'll get us off on something else. And I talk to her about the walk. I wish I could tell her about Jenny.

"Dad says if they don't change their minds and give you a proper welcome, why don't you just walk around Waseca?" We laugh for about ten minutes over that. Aside from the news about the toughs, it's the only mention either of us makes of Waseca.

When Mom leaves in the morning, it's arranged that she and Dad will pick me up Friday night at the junction of Routes 18 and 169 and drive me to Clear Lake for the weekend. A secret vacation from the walk that no one else will know about. She'll bring me back Monday morning, just the way she used to take me out to the woods when I was a kid; and on Saturday I'll walk into Waseca at noon, just the way I said I would; and the world walk will be over.

CHAPTER 24

THE ONLY PERSON OUTSIDE THE FAMILY WHO KNOWS THAT I am spending Saturday and Sunday at home is Dick Woodbury, a *Time* magazine correspondent. He is there, too, most of the time, and on Monday he drives back with me and we walk together part of the day. He kids me about how I come on to people. OK. I admit it; I like to shake people up. Jenny's told me that a thousand times. Even Jenny and I have had some pretty fierce fights. Doesn't mean a thing. I can be nice if I want to. Take that TV interview I gave in Fort Dodge, for instance. I explained to the guy very carefully that when I said that I'd met a lot of dumb foreigners, I meant it; they were dumb all right, but there were reasons for it, naturally. Sister Stella, in India, explained to me that many of the children in these so-called Third World countries die before they are seven years old, and of those who survive, only ten out of fifty develop normally. The rest suffer protein deficiencies and are listless, physically and mentally. As soon as I said all that the guy interviewing me lost that tense look he'd been wearing and smiled and nodded his head up and down; and they got a lot of calls after the program was over from people who said they understood now what I meant to say. "So why don't you always explain what you mean a little more carefully?" Dick says.

I laugh. " 'Cause I don't want to," I say. "It takes the fun out of it."

The next day two reporters from the *Mankato State College Reporter* walk with me for about fourteen miles. "How did you get the idea for the walk?" one of them asks, so I tell him. Every time I tell the story the whole scene flashes before me. I'm there.

January 14, 1970. I'd flicked off the lights in the projection booth, and walked downstairs to Rich's office, like I always did. He was checking the take with Marty, the cashier, so I went over to the refrigerator and helped myself to a beer.

"I could see that movie fifty times," I said, and plopped down into the brown vinyl armchair. "That Clint Eastwood. Christ!"

Later, as usual, Rich and I got talking and I told him about my idea to take a Jeep to the end of South America and back.

"It's been done," he said and yawned. "Why don't you walk around the world? No one's ever done that before."

It was so simple. You didn't need a lot of fancy equipment, like a sailboat or a balloon, or even any special clothes. All you had to have, basically, were your own two feet.

"How do you know it hasn't been done?" I asked him. I was already kind of trembling. He reached behind him, pulled out the *Guinness Book of World Records,* and showed me. Still, I could hardly believe it. Man had been to the moon, but he hadn't yet walked around his own planet. Brother!

He had it all figured out, too. He and his wife, Marge, would walk to New York City, build a raft, and take it to Europe; then walk to the other side of Russia, build another raft, and take that to California; and then walk back here to where they'd started. A raft? Right away we began arguing.

I didn't see the point of that. What were you going to do, walk in place all the way across? You couldn't walk on water, that was for sure, so what you meant when you said walk around the world was walk around the land part of it from ocean shore to ocean shore. It didn't matter how you got across the water. Besides, on a raft you could get drowned. No, I didn't see the point of that at all. We stayed up all night talking.

Funny. In a few days, I'll see Rich again. He and Marge and Pete are coming out to join me for the final day's walk into Waseca.

All along the way, now, people are greeting me. There's a group on the sidewalk in practically every little town. On Wednesday morning I cross the border into Minnesota at a place called Elmore and the whole town, practically, is out there. They've let the kids out of school and there're a bunch of newspapermen from all over the state. "How are you feeling about walking into Waseca?" someone asks me.

"Fine," I say. "I've decided to take a Christian attitude toward those people." That gets a lot of laughs. The media has been playing up those ridiculous resolutions that the chamber of commerce adopted about me. Half the crowd walks with me out of town when the press finishes asking questions and some of the older kids walk a couple of miles farther. All along the way, people stop or come down to the road and wish me well. The light shining in their eyes. That's what I love. I feel like reaching out and touching them. The next day, ten miles outside of Mankato, a girl named Jean parks her car by the side of the road and asks if she can walk with me. "Sure," I say. As we walk along more and more people join us until, by the time we reach the town, there must be fifty or so. The mayor of Mankato, Vernard Lundin, is at the city limits with a group of officials and well-wishers (and, of course, the press); and after he signs my scroll and officially welcomes me into the city, he and most of the others fall in beside me and we walk downtown together through cheering crowds. They put me up at a nice motel and then leave me alone to spend a quiet night. I'm just about to go to bed when there's a knock at the door, and I hear someone whispering and the sound of running feet. I grab the heavy walking stick I've been carrying for the last few days, in case these are the Waseca toughs I've been hearing so much about, and walk over to the door. "Who is it?" I shout. A girl's voice answers, and I open the door and there are three co-eds from the State College who just wanted to come down and meet me. Well, naturally, I let them in and we sit around for a while and talk about the walk, and when they leave I write each of them a little message on a sheet of motel stationery and sign it. It's fun. It tops off the day. Fans! Of course you love them.

I'm having breakfast the next morning, and in come Pete and Rich and Marge. We all hug and clap each other on the back, and before we're even sitting down we're talking away a mile a minute.

"So what makes you think it'll be better in St. Paul?" I ask.

"It's got to be. Besides, it's a bigger theater."

"Big enough for another ocelot?" Rich and I laugh. Marge doesn't. Rich is always getting these crazy ideas and losing his shirt. Once, he'd bought this real wild ocelot to promote a jungle movie he was showing. Used to take it to the theater with him and parade it all over town. Some mothers started complaining, and when he wouldn't get rid of it, they threatened to throw him out of town. It was plenty wild, too, and Rich had the scars to prove it.

Probably Marge is behind the move, I think. Actually, neither one of them is a small-town type, that's for sure.

"Thought you were going to behave, this time?" I say. "White shirt, tie, Rotary, the whole bit." Rich shakes his head. "The boys down at the chamber of commerce give you a hard time?"

"Let's just say we decided to give the big city a try," says Marge and lights a cigarette. Crazy Rich. Good old buddy. Probably spent more time with Rich and Marge that whole year before the walk than I did at home.

It's almost noon before we get going, what with the press crowding around and all the people. It sure is good to see Pete here. I thought for a while he might not come. He looks at me when I say this as if I must be kidding. Two Kunst brothers started the walk and two Kunst brothers are going to finish it. Isn't that what he said? Sure. Even though he had to skip the last three thousand miles. All the way out of town we can hardly make our way through the crowds. So many cars stop that the highway patrol finally has to escort us. We're walking slowly because there are so many people with us and anyway there's no rush now. I'm talking away with Rich or Pete or Marge and signing autographs, and every once in a while somebody comes up and tells me something that puts me even higher on cloud nine than I already am. Like the guy who's on his way from Idlewild, Michigan, to Waseca to honor the first man ever to walk around the world: Andy Horujko, who spent two years tramping from Alaska to the tip of South America.

While Waseca's been stewing over how to handle me, Janesville long ago decided to give me a real hero's welcome. We walk into town under a banner that reads "Welcome Dave Kunst—World Traveler." Not only is the mayor there and a huge crowd, but best of all, there's a band. It strikes up "It's a Long, Long Way to Tipperary," as we come into view, plays a few numbers, and ends with "Mule Train." I sing out in my gorilla voice and I can almost hear John singing, too. I'm crying a little as we walk up to the official welcoming committee. I can't help it. That bum. That son-of-a-bitch. He should *be* here. Everyone who can fit in goes inside the town hall and Mayor Hendricks makes a long speech, interrupted every few minutes by claps and shouts, about what a great feat it is I've accomplished. Fifteen thousand miles. Twenty-two pairs of shoes. Four years, three months, and sixteen days. Over twenty million steps. "Only a Minnesotian could have done it," he yells, and the crowd cheers. "My only regret is that Dave Kunst didn't come from Janesville." Bedlam!

That night, at a little celebration over at the mayor's house, I get a phone call from Chicago. It's Dick Woodbury. My story's gotten almost a page in the Nation section of *Time* magazine. And there's a full-length picture. It takes me awhile to get to sleep that night, believe me.

Saturday turns out to be cloudy, and everybody's worried that it's going to rain. It was raining when we left town, four and a half years ago, so I don't much care if it rains when I come back. It would be fitting, in a way; though naturally, I hope it won't. Hundreds of people are walking with us now: newspapermen, TV people, all kinds of friends and well-wishers. I'm kind of out of it all, though. In a couple of hours the walk'll be over, and like before you die your whole life passes in front of you, so now, at the end of the walk, the last four and a half years are rolling over and over in my mind.

John: "Wait up, can't you?" I stop and up he hobbles. And then he's lying on the ground, curled up as if asleep. And I turn him over and he turns too easily. There's nothing behind his face but his bloody brains splattered all over the dirt. John! You hear me? John! So many things, so many places. So many different people. "You're a cobber, you are. A fair cobber!" her laugh, her bray, as I called it, just to tease her. She and John would have liked each other. She and John and Pete and me: the four Musketeers. Dad with his cigar. That twinkle he gets sometimes. Walk *around* the town. Not a bad idea.

I sign the autographs and smile and talk to people steadily, it seems; but I'm really up there above myself, looking down and smiling at all this. The walk's over, just about, and part of me wants not just to walk around Waseca but keep going for another round. To think of not walking anymore. It seems so strange. If it weren't for Jenny, the idea would even be tempting: "Kunst Refuses To End Walk. Will Go Around Again." "Kunst Won't Cross Finish Line. Will Walk Around the World the Other Way." Jenny, oh, Jenny! Maybe someday we'll walk around together.

Off in the distance I see a crowd gathered. Figures rush out toward me. It's a group of friends and relatives from Caledonia, in the southeastern part of the state where I was born. The sides of the highway are lined now with cars and people. We pass the sign that says "Jesus Loves Me" with "And You Too Dave!" written underneath it, and some people cheer and some boo and it's all for me. The dumb city fathers, everyone's laughing at them. And suddenly it hits me why we went on the walk to begin with. Ever since we started, people have been asking us that question. Depending on how we felt, we'd say "the adventure of it," or "the challenge," or "for UNICEF," or "because it's such a great learning experience." But now I know that none of those reasons is the real one. The real reason is to do something that's never been done, to be the first! To take something that seems impossible and to go ahead and do it, no matter what. If you can walk around the world, then you can do anything. And so can anyone.

Up ahead are the city limits of Waseca and there's Dad holding Willie Make It. When all the fuss started, they canceled all that, but I'm glad to see they've had a change of heart. Good old Willie. "You start the walk with a mule. You end the walk with a mule." I hear Pete saying it before he opens his mouth. The boys are there. I lift them up and hug them. Dad joins the crowd and Bradley and Danny walk just behind Pete, who's at Willie's tail now, just where John used to be. Mom and Debra and Jan and the welcoming committee are waiting for me at the State Theater where it all began. That last mile is wild. People waving and running out to greet me practically every step of the way. Willie's the same old mule, just four years older; and sure enough she slobbers all over my right arm the way she always did, and Pete has to touch her on the ass every once in a while to keep her pace steady. Through all the noise and shouting and the people yelling—some I recognize, others I've never seen before—I hear John's laugh. It's bellying out there behind me. He's with us. Sure he is. And he's loving every minute of it, too. "Yahoo!" I yell, and he yells back. "Nobody knows and nobody cares." Is he saying that? Wrong, old buddy! "Everybody knows and everybody cares!" Sing it out! Again!

And then we're rounding the last turn before we actually get onto State Street and head for the theater, and the church bells start ringing. The sky clears for a moment and the sun comes out. I know it sounds corny but it's what happens. The sun hasn't been out all day. And it picks that moment to shine its blessing on us. Well, what else can you call it but a sign! We turn the last corner and the whole roped-off street is packed with cheering people. There must be over a thousand of them. Right around the marquee where they've set up the welcoming committee there are so many TV cameras and news people I don't see how there's room for us. There's Robert T. Smith. He's going to be the master of ceremonies. No Mayor Swanson, but he's about the only one who's absent. Things quiet down pretty quickly, considering what a crowd there is. I don't remember everything that's said. What I mostly latch on to is old Robert T.'s opening words. After that I'm sort of laughing and crying too much to concentrate on much else. "Waseca is officially honoring the World Walk. But some, including me, are here to welcome Dave Kunst."

Later on, I place my foot on the spot where John and I started the walk on Saturday, June 20, 1970, and officially become the first man to walk around the world. "Speech! Speech!" someone yells, and almost right away there's silence. "I just walked around the world and I feel great," I say, and I wave my hat and the crowd cheers and shouts hurray.